BUILDING CONTRACT
ADMINISTRATION AND PRACTICE

CONSTRUCTION TECHNOLOGY AND MANAGEMENT

A series published in association with the Chartered Institute of Building.

This series covers every important aspect of construction. It is of particular relevance to the needs of students taking the CIOB Member Examinations, Parts 1 and 2, but will also be suitable for degree courses, other professional examinations, and practitioners in building, architecture, surveying and related fields.

Project Evaluation and Development
Alexander Rougvie

Practical Building Law
Margaret Wilkie with Richard Howells

Building Technology (3 volumes)
Ian Chandler
   Vol 1 Site Organisation and Method
   Vol 2 Performance
   Vol 3 Design, Production and Maintenance

The Economics of the Construction Industry
Geoffrey Briscoe

Construction Management (2 volumes)
Robert Newcombe, David Langford and Richard Fellows
   Vol 1 Organisation Systems
   Vol 2 Management Systems

BUILDING CONTRACT ADMINISTRATION AND PRACTICE

James Franks

B T Batsford Ltd · *London*

in association with the Chartered Institute of Building

© James Franks 1991

First published 1991

All rights reserved. No part of this publication may be reproduced, in any form or by any means, without permission from the Publisher

Typeset by Best-set Typesetter Ltd, Hong Kong
and printed in Great Britain by
Courier International Ltd, East Kilbride
Scotland

Published by B T Batsford Ltd
4 Fitzhardinge Street, London W1H 0AH

A catalogue record for this book is available from the British Library

ISBN 0 7134 6759 2

CONTENTS

List of illustrations   x

Abbreviations used in the text   xi

Acknowledgment   xii

Introduction   xiii

Section 1   CONTRACT PRINCIPLES, PROCEDURES AND PRACTICE   1

Section 1A   PRECONTRACT PROCEDURES   1

1.1   **Background**   1
   Selection of procurement path and form of contract   1
   RIBA plan of work   3
   Inception and feasibility (stages A and B)   3
      The brief   3
      Selection of design and management team   5
   Outline proposals (stage C)   6
      Tasks to be undertaken   6
   Scheme and detailed design information (stage D to F)   7
      Tasks to be undertaken   7

1.2   **Cost planning framework**   8
   The objectives   8
      Cost planning data   8
      Units of cost   8
      Sources of data   10
   Cost planning practice – inception to detailed designs (Stages A to E)   10

1.3   **Estimating and tendering**   11
   Invitation to tender   11
   Cover prices   11
   Decision to tender   11
   A methodical approach to estimating   12
   Data for estimating   12
   Overhead costs   12
   Profit margin – the commercial decision   15
   Break-even calculations   17
   Submission of tender   18
   The CIOB estimating and tendering practice   18
   Cost data cycle   23

Section 1B   MOBILISATION OF THE WORKS   24

1.4   **Sub-contractor selection**   24
   Precontract negotiations   24
   Selection   24
   Factors affecting selection   24
   Assessment procedures   25

1.5   **Entering into contract**   26
   Project management tasks   26
   Aspects of contract   26
   Express and implied terms   27
   Privity of contract   28
   Assignment   28
   Standard forms of contract   28

1.6   **Nominated, named and domestic sub-contractors**   29
   Alternative methods of selection   29
   Contractor's agreement to nominated and/or named sub-contractors   30
   Nominated procedures   30

1.7   **Mobilisation**   41
   By the project manager   41
   By the contractor   42
   Uniform procedures   42

Section 1C   EXECUTION OF THE WORKS   42

1.8   **Carrying out the works**   42
   The contractor's responsibility and obligations   43

Contract sum 44
Contract sum adjustments 44
Errors in contract bills and discrepancies between documents 44
Architect's instructions 44
Documents and certificates 46
Administration of documents 46
Co-ordinated project information 46
Custody of contract documents 48
Descriptive schedules and master programme 48
Variations 49
Provisional sums 49
Valuation of variations 49
Valuation rules 49
Administering variations 50
Identifying variations 50
Recording variations 50
Statutory obligations 51
Positioning the building/s 52
Materials, goods and workmanship 52
Exclusion of persons by architect 53
Royalties and patent rights 53
Access for architect to the works 53
The contractor's 'person-in-charge' 53
The clerk to works 53
Injury to persons and property 54
Interim certificates and payments 54
Taxation 55
Works by employer or his employees 55
Group decisions and communications 56
Meetings 56
Group management methods applied to construction contracts 57
Sub-contract administration 59

Section 1D PRACTICAL COMPLETION OF THE WORKS 62

1.9 Practical completion and final payment 62
Practical completion checklist 62
Practical completion 63
Partial possession by employer 64
Final account 64
Effect of final certificate 65

1.10 Alternative methods of terminating the contract 65
Determination by employer 65
Determination by contractor 66
Determination by employer or contractor 66
Outbreak of hostilities and war damage 66

Section 1E CONSTRUCTION PROJECT COST MONITORING AND CONTROL 67

1.11 Elemental cost plan as basis for expenditure monitoring 67
Integrated cost management 70
Corporate organisation structure 70
Budgetary control, principles and benefits 71
Budgetary responsibility 71
Construction project profit centres and budgets 71
Computer applications to budgetary management 71
Cost:value reconciliation of work in progress 75
Reconciling the cost: value of operations 75
Recording operational costs 76

1.12 Case study 78
Scenario 78
Situation 78

1.13 Exercises 81

1.14 Sources and further reading 83

1.15 References 83

Appendices – extracts from procedures manuals 85
(a) Index to procedures manual
(b) Materials purchasing procedures
(c) Schedule of purchasing forms

CONTENTS

Section 2   DELAYS, DISRUPTION AND REIMBURSEMENT   89

2.1   **Site possession by contractor, completion and postponement**   89
Liquidated damages   89
Relevant events   89
Delay   90
Written extension of time   90
Action by architect within 12 weeks   90
Sub-contractors' obligations   90
Earlier completion date   91
Shortened period for completion   93
Relevant events   93
Antiquities   97

2.2   **Loss and expense**   98
List of matters   98

2.3   **Claims for loss and expense**   99
Preparing a claim   100
Selection of matters   100
Recording the facts   100
Examining the facts   101
Analysis or diagnosis   102
Format of claim   104
Contract particulars   104
Statement of events leading to claim   104
Management involvement   105
Statement of claim   105
   Global approach or detailed statement?   106
   Cases supporting the global approach   106
   Calculations for a detailed statement   106
   Management and supervisory costs   106
   Supervisory visits   107
   Labour and plant costs   107
   Materials abstract   108
   Cost differences   109
   NEDO formula   109
Claims against sub-contractors, suppliers and insurers   109
Site supervision and management costs   109
Establishment/head office costs   110
   Formulae   110
   Cases relating to establishment costs   111
Finance charges   112
Profit   112
Cost of preparing the claim   112
Presentation of claim   113
Who prepares claims?   114
Credibility of claim   115

2.4   **Role of the expert witness**   116
Briefing   116
Preliminary report   116
Format of report   116
Consultation with other experts   117
Exchange of reports   117
Meeting of experts   117
Expert's integrity   117

2.5   **Case studies**   117

2.6   **Exercises**   118

2.7   **Sources and further reading**   119

2.8   **References**   119

Section 3   INSURANCE   121
(*by Brian Thornton*)

3.1   **Background**   121

3.2   **Principles**   121
Utmost good faith   121
Proximate cause   121
Insurable interest   122

3.3   **Main types of insurance**   122
Practice   122

3.4   **Loss or damage to material property**   122

3.5   **Legal liability**   123
Employer's liability   123
Public liability   124

3.6   **Professional indemnity**   124

3.7   **Motor insurance**   125

3.8 **Contract guarantee bonds** 125

3.9 **Premium** 126

3.10 **Self-insurance** 127

3.11 **Claims** 127

3.12 **Exercises** 128

3.13 **Sources and further reading** 129

Section 4 RESOLUTION OF DISPUTES 131

4.1 **Introduction** 131

4.2 **Disputes and differences – alternative approaches for resolution** 132
Conciliation 132
Mediation 132
Mini-trial 132
Adjudication 133
Litigation 133
Arbitration 133
Which approach to use? 133

4.3 **Arbitration – the legal framework** 134

4.4 **Arbitration as a standard condition of contract** 137

4.5 **Four procedures in outline** 138
The traditional process 138
The procedures in the rules 138
Procedure without hearing 140
Full procedure with hearing 141

4.6 **Procedures leading to hearing – in more detail** 142
The dispute, agreement to submit to arbitration and appointment of arbitrator (steps 1–3) 142
The arbitration agreement 142
Reference to arbitration 142
Opening the reference 143
Appointment of arbitrator – by agreement or nomination? 143
The arbitrator's qualifications, expertise and skills 144
Powers of arbitrator 145
Joint arbitration 146
Bringing the dispute to court 147
Legal representation 147
Service of documents 148
Preliminary meeting 148
Directions 149
The Scott Schedule 151
Pleadings 152
  Points of claim 152
  Further and better particulars 152
  Counterclaim, defence, etc 152
Examination of property by parties 152
Inspection by arbitrator 152
Discovery 153
Privileged documents 153
Agreement of figures and facts 154
Evidence 154
  Documentary 154
  Oral 154
  Real 154
  Of fact 154
  As to opinion 154
  Hearsay 155
  Burden of proof 155
  Preponderance of probability 155
  The admissibility of evidence 155
Offers to settle 155

4.7 **The hearing** 159

4.8 **The award** 159
Enforcing the award 160
Remission and setting aside award 161

4.9 **Arbitrator's fees and expenses – costs** 162

4.10 **Advantages and disadvantages of arbitration** 163

4.11 **Case study** 163
Arbitration, Lucifer v Barchester
Synopsis 163
The hearing 164
Closing addresses 170
The award 173

4.12 Exercises 174

4.13 Sources and further reading 174

4.14 References 175

Section 5 ALTERNATIVE CONTRACTUAL ARRANGEMENTS 177

5.1 Background to change 177
The Emmerson and Banwell Reports 177

5.2 Procurement systems: types and terms 178
Designer-led competitive tender 179
Traditional system 179
Fast-track 182
Designer-led construction works managed for a fee 182
Two-stage tendering 182
Management contracting/construction management 182
Package deal, design-and-build 185
Project manager/client's representative 189
The British Property Federation System 191

5.3 Project time: cost relationships 196
Case study illustrating the relationship between time and cost 196

5.4 Comparison of procurement systems 198
Client's need 198
Traditional system 198
Management for a fee 199
Package deal, design-and-build 200
Project management/client's representative 200
Conclusions 201

5.5 Case study 201

5.6 Exercises 201

5.7 Glossary 202

5.8 Sources 204

5.9 References 204

Index 205

List of illustrations

1 **Contract principles, procedures and practice**
1.1 Simple guide for selecting the appropriate JCT form of contract
1.2 Flow chart for selecting the appropriate JCT form of contract
1.3 RIBA plan of work
1.4 Cost planning – advice, costing units and data sources
1.5 Calculating gross floor area
1.6 Calculating the volume
1.7 Data context for pre-tender procedures
1.8 Pre-tender data flow
1.9 The intuitive approach to profit margin calculation
1.10 Record of tenders
1.11 Tendering efficiency calculation
1.12 Break-even analysis
1.13 Decision to tender – activities
1.14 Project appreciation – activities
1.15 Estimator's report and adjudication – activities
1.16 Production and cost data cycle for construction projects
1.17 Contractual relationships between parties
1.18 Procedure using Basic Method
1.19 Form of agreement NSC/1
1.20 Conditions for agreement NSC/1, page 9
1.21 Conditions for agreement NSC/1, page 10
1.22 Form of nomination NSC/3
1.23 The effect of drawings arrangement
1.24 Design variation sheet
1.25 Aims of meeting
1.26 Checklist for meeting convenor
1.27 Elemental cost plan for swimming pool
1.28 Costed network for swimming pool
1.29 Costed bar chart for swimming pool
1.30 Expenditure forecast for swimming pool
1.31 Properties of TASC, Generation of forecast for given values of dependent variables
1.32 TASC dependent variables
1.33 TASC distortion module
1.34 TASC Kurtosis variables
1.35 Periodic and cumulative values
1.36 Cost:value reconciliation standard reporting form
1.37 Daily allocation sheet (for labour and cost allocation)

2 **Delays, disruption and reimbursement**
2.1 Application for instructions, drawings, etc
2.2 Recording revisions of drawings
2.3 Expenditure profiles

4 **Arbitration and resolution of disputes**
4.1 Structure and relationship of English courts relating to civil disputes
4.2 Flow chart for arbitration
4.3 Procedure without hearing – the timetable
4.4 The Scott schedule of items in dispute

5 **Alternative contractual arrangements**
5.1 Alternative systems for building procurement
5.2 The traditional system
5.3 Management (fee) contract system using two-tier tendering approach
5.4 Design-and-build system
5.5 Relationship between parties for project management systems
5.6 Project management
5.7 BPF system for building design and construction
5.8 The pre-construction timetable
5.9 Time:cost case study for commercial project
5.10 Rating the systems

Abbreviations used in the text

ACA Association of Consultant Architects
ACAS The Advisory, Conciliation and Arbitration Service (created by the Employment Protection Act 1975 to promote improvement of industrial relations)
ACE Association of Consulting Engineers
BCIS Building Cost Information Service (of the RICS)
BEC Building Employers' Confederation
BPF British Property Federation
BPIC Building Project Information Committee, comprising representatives of the ACE, BEC, RIBA and RICS
CCPI Co-ordinating Committee for Project Information sponsored by RIBA, RICS, ACE and BEC in 1979. The committee's objective is to bring about a general improvement in the documents used for the procurement and construction of buildings
CPI Co-ordinated Project Information
CD 81 JCT standard form of building contract with contractor's design 1981 edition
CIArb Chartered Institute of Arbitrators
CICA Construction Industry Computing Association
CIOB Chartered Institute of Building
DoE Department of the Environment
DOM/1 Standard form of sub-contract for domestic sub-contractors published by BEC which is appropriate for use where the main form of contract is some of the JCT standard forms
ICA Institute of Chartered Accountants
ICE Institution of Civil Engineers
IFC 84 JCT intermediate form of building contract for works of simple content 1984 edition
JCT Joint Contracts Tribunal
JCT 80 JCT standard form of building contract, 1980 edition (the 'with quantities' version has been used for this text)
NEDO National Economic Development Office
NSC/1 JCT standard form of nominated sub-contract tender and agreement
NSC/2 JCT standard form of employer/nominated sub-contractor agreement
NSC/2a Agreement NSC/2 adapted for use where NSC/1 has not been used
NSC/3 JCT standard form or nomination of a sub-contractor where NSC/1 has been used
NSC/4 JCT standard form of sub-contract for sub-contractors who have tendered on Tender NSC/1 and executed Agreement NSC/2 and been nominated by Nomination NSC/3
NSC/4a NSC/4 adapted for use where NSC/1 has not been used
OED Oxford English Dictionary
RIBA Royal Institute of British Architects
RICS Royal Institution of Chartered Surveyors
SMM7 Standard Method of Measurement, 7th edition
SSAP Statement of Standard Accounting Practice published by the ICA
UNICITRAL The United Nations Commission on International Trade Law

Acknowledgment

Thanks and acknowledgment to the following people who have contributed to the production of this text. Their names are mentioned, generally, in the order that their contributions appear.

To Oliver Longley and John Ebdon, James Longley & Co, and to John Kirby, IDC for allowing me access to their companies' procedures manuals which have influenced the 'contract principles, procedures and practice' section.

To colleague Farzad Khosrowshahi whose contribution on the operation of his Advanced 'S' Curve (TASC) provides an example of computer application to cost monitoring and control.

To Brian Thornton, Willis Wrightson Construction Risks who contributed the section on insurance.

To Harold Crowter whose comments and suggestions on the 'resolution of disputes' section were gratefully incorporated and to Norman Royce, doyen of building arbitration who kindly allowed me to include his practice arbitration as the case study for this section.

To Julian Vickery, Greycoat who provided data for the time: cost case study in the 'alternative contractual arrangements' section which demonstrates the advantages of early completion on commercial projects brought about by fast-track projects.

To colleague Julia Lemessany who pointed me towards significant findings in the NEDO faster building for industry and commerce reports, (with which she was closely involved), and to Ron Denny, formerly deputy director BPF for providing an available ear.

To Peter Harlow, CIOB Information Resource Centre and colleague David Coles who advised on aspects of the CIOB syllabuses in relation to the text and made valuable recommendations on the draft, to other colleagues at South Bank Polytechnic who assisted with suggestions, word processing, graphics and numerous other production matters and to Rod Howes who 'enabled' them.

To Geoff Heathcote who patiently proofread, and to others whose contributions may not have been acknowledged above, apologies for my oversight and sincere thanks.

To my wife who at times has been relegated to the role of 'book widow', thank you.

James Franks
Department of Construction Management
South Bank Polytechnic, London, 1991

INTRODUCTION

This text is intended for use by construction management and surveying students in general and those preparing for the Contract Administration examination as part of the Membership Part II and the Contract Administration and Practice paper of the Direct Membership Examinations of the CIOB in particular. It takes note of the syllabuses and commentaries on the syllabuses.

The syllabus structures for the two examinations are essentially the same. This text adopts the section format and weighting shown in the syllabus for the Membership Examination Part II, namely:

Section 1   30% Contract principles, procedures and practice
Section 2   20% Delays, disruption and reimbursement
Section 3   10% Insurance
Section 4   20% Arbitration (and resolution of disputes generally)
Section 5   20% Alternative contractual arrangements

The scope of the syllabuses is wide and cannot be covered in depth in a text such as this which must necessarily be restricted to providing an introduction to all the subjects. Case studies, exercises and sources for further reading are provided at the conclusion of each section.

The CIOB syllabuses contain references to approximate estimates, cost planning and control, tendering procedures and other matters which are dealt with in considerable depth in other CIOB syllabuses. Such matters are discussed in this text only in sufficient depth to demonstrate the principles, procedures and practices involved in their administration.

Unless otherwise stated, the JCT Standard Forms of Building Contract have been used to demonstrate practice and procedures.

Frequent revisions to JCT publications make it difficult to ensure that the clause numbers given in the text are current. The reader should use the most recent editions.

Computing technology develops so rapidly that reference to specific programs and hardware has been kept to a minimum despite its obvious application to the administration of contracts. The CICA should be consulted for 'state of the art' advice on current developments. An example of computer application is provided in section 1E where a mathematical-based model for forecasting construction project expenditure/income patterns is explained in some detail.

Client, customer or employer? The party to the contract who commissions the works and typically pays for them is usually referred to by the other parties as 'the client' although the NEDO-sponsored 'Faster building...' reports refer to him as 'the customer'. His title in JCT contracts is 'the employer'. In this text he is referred to as 'the client' but in the context of JCT conditions of contract *employer* seems appropriate and has been used. *Customer* has been used in quotations from the NEDO reports.

Repetition There is some repetition of text between sections. This is intentional. It is expected that the sections will be read separately to some extent at least.

Definitions A *contract* is defined as: 'a mutual agreement' which is 'enforceable by law' and *to administer* as: 'to manage (as a steward), to carry on, or execute (an office, affairs, etc), to manage the affairs of (an institution, town, etc)' (OED).

Contract administration in the building industry is concerned with providing a framework (the appropriate information, communications, legal and contractual environment) which will enable the managers to manage the firm's contracts.

The aim of administration is to establish the principles which the manager applies in practice and the procedures which he follows. The procedures should act as *enablers* rather than *inhibitors*. They should endorse the responsibility and authority of managers so that all concerned may perform their tasks to achieve the corporate and project aims.

For the sake of consistency with the CIOB syllabus the term *contract administration* is used in this text but 'project' is used increasingly in preference to 'contract' from which it follows that it would be equally appropriate to refer to *project administration*.

The contractor's changing role Traditionally the contractor has been concerned only with 'building', with, to use the words of many JCT standard forms of building contract, carrying out and completing the works designed by others. He has had no responsibility for design.

In recent years that role has been changing. Some contractors have widened their activities to include designing in addition to building. Such design-and-build firms have responsibilities to the environment in addition to their responsibilities to their client.

Other contractual arrangements have evolved for the procurement of projects which determine the ways in which they must be administered. An appreciation of these procurement systems is essential when establishing an administrative framework.

Section 5 explains the background, operation and characteristics of the alternative systems.

Section 1
CONTRACT PRINCIPLES, PROCEDURES AND PRACTICE

Section 1 is concerned with precontract procedures from inception to completion of the contract works. The subject of the section has been subdivided to facilitate study:

Section 1A **Precontract procedures**
concerned with:
selection of the appropriate form of contract
cost planning, related to the RIBA plan of work
estimating and tendering for work

Section 1B **Mobilisation of the works**
concerned with:
sub-contractor selection
entering into contract

Section 1C **Execution of the works**
using JCT 80 conditions as the model

Section 1D **Practical completion of the works**
concerned with:
partial possession and practical completion
final account procedures
alternative methods of terminating the works

Section 1E **Construction project cost monitoring and control**
concerned with:
use of elemental cost plan as basis for expenditure monitoring
budgetary control principles
computer applications to budgetary management
cost: value reconciliation and reporting

Section 1A
PRECONTRACT PROCEDURES

1.1 Background

At each stage in the development of a project, decisions must be made to ensure that the client's requirements regarding the fitness of the building for his purpose, its cost and delivery time are met.

At each stage tasks have to be undertaken. A report on management for commercial projects[1] identified more than seventy separate tasks which are considered here, broadly, under the headings identified in the RIBA plan of work to which reference is made on page 3.

This section, section 1, is concerned with establishing the framework within which a project is administered. It is not concerned with detailed descriptions of, for example, methods of construction planning or with cost planning and control, etc, which are outside the scope of this text.

Selection of procurement path and form of contract Selection of the most appropriate procurement path is one of, if not the most important decision to be made by the client who intends to build. The decision is so fundamental that section 5 is devoted to

CONTRACT PRINCIPLES, PROCEDURES AND PRACTICE

Designer of the Works	Contract Documents	Value Range	Recommended Form of Contract		Comments	Type of Contract	Fluctuations available	Subcontract forms		Comments
			Private Sector	Public Sector				Nominated	Domestic	
Architect or other Professional engaged by Employer	Drawings and full Bill of Quantities	£250,000 (1980 Prices)	JCT 80 Private with Quantities (80/PW)	JCT 80 Local Authorities with Quantities (80/LAW)	For general use and where works are of a complex nature or with a high content of building services or other specialist work or exceed 12 months duration.	Lump Sum	Tax; Full; Formula.	NSC/4 or NSC/4a	DOM/1	NSC/4 used where S/C has tendered on NSC/1, executed NSC/2 and been nominated on NSC/3 under the Basic Method. NSC/4a used where S/C is Nominated under the Alternative Method.
Ditto	Drawings and Specifications	Up to £100,000 (1980 Prices)	JCT 80 Private without Quantities (80/PWO)	JCT 80 Local Authorities without Quantities (80/LAWO)	Not recommended if works are of a complex nature that can best be explained to the tendering Contractor by use of a Bill of Quantities.	Lump Sum	Tax; Full; Formula.	NSC/4 or NSC/4a	DOM/1	Ditto
Ditto	Drawings and Approximate Bill of Quantities	General	JCT 80 Private with approx. Quantities (80/PWA)	JCT 80 Local Authorities with approx. Quantities (80/LAA)	For use where extent of work is not fully known (eg alterations or repairs) or where an early start is required before detailed contract documents can be prepared.	Re-measure	Full; Formula.	NSC/4 or NSC/4a	DOM/1	Ditto
Ditto	Drawings and Specifications or Bill of Quantities	up to £250,000 (1984 Prices)	JCT Intermediate Form of Building Contract (IFC/84)		For simple contracts not exceeding 12 months duration without high content of building services or other works of a specialist nature.	Lump Sum	Tax; Formula.	N/A	NAM/SC or IN/SC	NAM/SC is compulsory where S/C has been 'named' by Architect. IN/SC is for use where S/C appointed by Contractor.
Ditto	Drawings and Specifications or Schedules	up to £50,000 (1981 Prices)	JCT Agreement for Minor Building Works (80/MW)		For simple contracts of short duration.	Lump Sum	Tax only.	N/A	N/A	
Ditto	Specifications with or without Drawings	General	JCT Fixed Fee Form of Prime Cost Contract		For use where nature or extent of works cannot be fully described or where an early start is required before design is finalised.	Actual Cost plus Fixed lump sum Fee	N/A			
Contractor	Employers Requirements, Contractors Proposals, Contract Sum Analysis	General	JCT Standard Form with Contractors Design (81/CD)			Lump Sum		N/A	DOM/2	
Part Architect or other Professional engaged by Employer and part Contractor	Drawings and Bill of Quantities plus Employers Requirements for the Contractor Designed Portion.		JCT Private with Quantities (80/PW) plus Contractors Designed Portion Supplement (81/CD/DP)	JCT 80 Local Authorities with Quantities (80/LAW) plus Contractors Design Portion Contractor (81/CD/DP)	For use with JCT-80 Standard Forms where part of the works (eg Piling, suspended floors, roof trusses, etc) are to be designed by the Contractor	Lump Sum	Tax; Full; Formula.	NSC/4 or NSC/4a	DOM/1 or DOM/2	

1.1 Simple guide for selecting the appropriate JCT form of contract **BEC, 1987** describing alternate contractual arrangements for the procurement of buildings. The JCT has produced standard forms of building contract for many of the options.

A table and flow diagram are shown as figures 1.1 and 1.2. These provide a simple guide for selecting the appropriate JCT form of contract.

When BEC published its guide and flow chart the only provision for management contracts was the JCT Fixed Fee Form of Prime Cost Contract. In 1987 JCT published the Management Contract which has much in common with the Fixed Fee Form but makes greater provision for sub-contracts through the Works Contract Conditions. In most respects the guide and flow chart for the Fixed Fee Form are appropriate for the Management Contract and Works Contract Conditions.

Generally, this section considers procedures from the standpoint of the contractor who undertakes a project adopting a traditional contractual arrangement. However, as he is increasingly involved in other contractual arrangements, such as design-and-build, a wider view has been taken where it is considered appropriate.

RIBA plan of work Figure 1.3 illustrates the stages in the RIBA plan of work. Figure 1.4 indicates the methods of cost estimation which are available at each stage. The sequence shown is applicable to projects following the 'traditional' client-architect-contractor selected by competitive tender contractual arrangement. Some of the stages may run concurrently, rather than sequentially, when the various fast-track contractual arrangements, discussed in section 5, are adopted. The builder's contribution at each stage depends on the procurement path used for a particular project.

Inception and feasibility (stages A and B) These stages are concerned with determining the brief, appointing consultants and deciding if the project is feasible.

The brief is a statement of the client's requirements. It may be in the form of a single sentence performance specification such as; 'design and build on a site owned by the client an office building which will provide the maximum return on his investment', or it may be a detailed specification of the materials to be used, accommodation required, architectural style, etc. Whichever form the brief takes, there is ample evidence that a well-stated brief is an essential ingredient of a successful project.

The checklist of 'data for inclusion in the brief' prepared by a local authority 'client' who regularly commissions residential building works includes:

> the title, address, telephone and Fax numbers of the project office and similar information regarding the client, project manager and residents affected by the proposed development.

The names of people concerned, including representatives of the residents, are listed.

Particulars of the project include:

- limits of site (boundaries, floor levels involved and party wall ownership)
- nature and scope of the works, indicative cost limits and time scale
- regulatory/planning/occupancy/etc, status (building regulations, extent of planning approval, extent of existing occupancy and of occupancy during the course of the works)
- data to be provided by the client, (drawings, existing reports, standard documentation)
- conditions to be used for contracts with consultants, contractors, sub-contractors and others
- reporting arrangements, (frequency, format, and project manager's expectations in respect of oral and written reports)
- meetings for progress monitoring, including arrangements regarding chair and recording
- residents' consultations (extent of consultants' responsibilities and project manager's expectations)
- consultants' terms of reference (their responsibilities, authority, extent of their

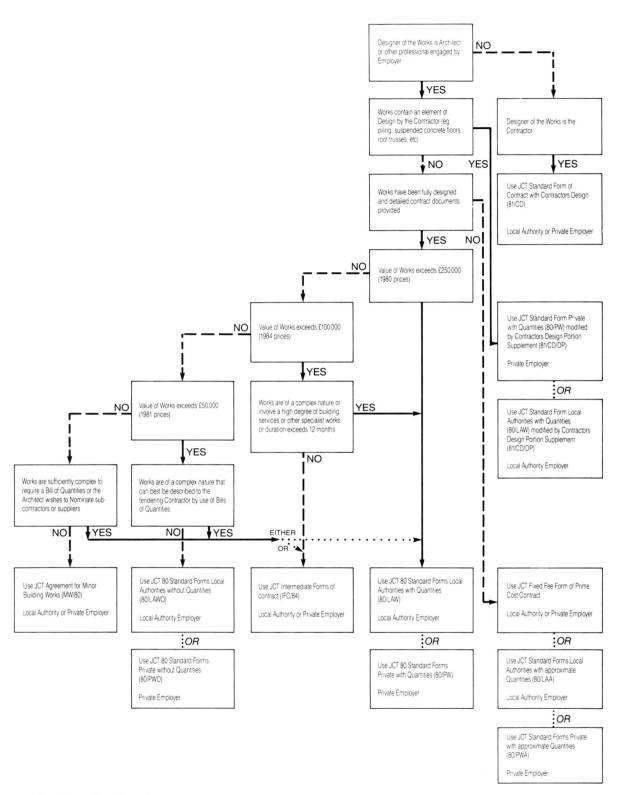

1.2 Flow chart for selecting the appropriate JCT form of contract BEC, 1987

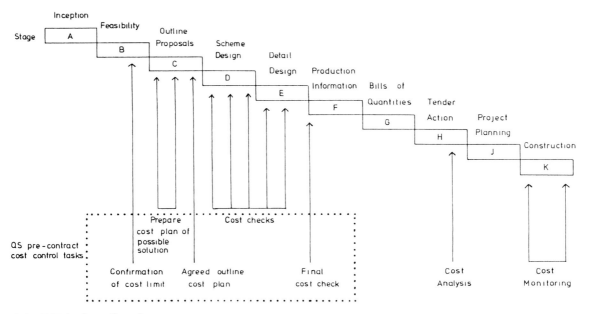

1.3 *RIBA plan of work*
    Building Bulletin 4: Cost Study (DES)

work, other consultants to be engaged, demarcation lines between consultants and conditions of employment).

The brief for a major project is seldom prepared without the assistance of experts, and because few projects proceed without regard to cost there is seldom a clearly defined interface between the inception and feasibility stages. The brief is shaped to a greater or lesser extent by the funds available for its realisation. Conversely, the brief determines the design and project cost.

Whether or not the consultants appointed to prepare the brief are subsequently appointed to be the design and management team depends on the nature of the project.

Selection of design and management team
For major projects, a project manager or client's representative will, typically, assist the client with selection of the design and management team. For smaller projects the client may decide on one of the procurement paths discussed in section 5.

Whatever the contractual arrangement the tasks to be carried out during the inception and feasibility stages include:

— selection of the most appropriate procurement path for the project
— selection of the design and management team
— market research (for a commercial project)
— considering site acquisition alternatives
— ascertaining town planning potential
— risk analysis
— viability studies
— considering effects of taxation, legislation, central or local government incentives in the form of grants, rate relief, etc
— ascertaining sources of finance
— estimating construction cost and delivery time
— determining the client's cost ceiling.

The above list is merely indicative. Checklists of tasks prepared by regular clients of the building industry vary considerably depending on the type of enterprise and 'product' with which they are concerned. For example, a checklist prepared for use by the project managers in the housing department of a local authority during the initial 'briefing consultants' stage lists his tasks as:

- collect and collate data for brief
- inspect site
- ascertain freehold/leasehold particulars
- notify interested parties
    - residents
    - planners
    - legal department
    - valuer
    - others
- cost aspects
    - ascertain out-turn budget
    - ascertain DoE allowances
    - decide amount to allocate for consultant's fees
    - decide amount to allocate for construction
    - decide amount to allocate for 'unforeseeable' items during construction
    - decide amount to allocate for other items
    - estimate cost of 'market factor' increases during total project period
    - estimate 'inflation factor' increases during total project period
    - determine sum included in consultant's briefs for construction cost
    - advise consultant if last item is present or out-turn budget expenditure programme
- determine indicative time scale and programme
    - design commencement date
    - pre-construction design period
    - construction commencement date
    - construction period
    - project completion date
- determine quality standards in relation to building life-cycle requirements
- check procedures with standing orders
- confirm brief with 'client'
- determine client interim financial review arrangements
- prepare terms of reference for consultants.

At the end of the feasibility stage the client and those concerned with realising the project should be satisfied:

(a) that the brief expresses the client's requirements and that the finished building will suit his purpose
(b) that the building can be realised within his budget
(c) that the building can be delivered on time.

Principles and procedures must be established accordingly.

Outline proposals (stage C) Having in stages A and B determined the client's requirements, stage C is concerned with making outline proposals to meet the requirements.

The decisions to be made may be fundamental. They may, for example, be concerned with decisions regarding the selection of site location or, if the site has been determined, with the 'mix' of building types (industrial, commercial, residential, etc) to meet the brief.

If the brief is concerned with provision of power to meet the needs of a specified community, decisions will be made regarding the selection of the most appropriate energy source.

Design proposals for buildings with alternative configurations or alternative components will be developed, costed and considered to determine if they are capable of being constructed within the client's time scale and budget.

Tasks to be undertaken during the outline proposals stage include:

- investigating site restrictions
- considering initial design proposals regarding structure, accommodation, layout and ratio of gross to nett areas of accommodation
- considering feasibility studies
- consulting design team to produce optimum scheme in design, cost and profitability terms
- agreeing in principle proposed scheme with client.

The local authority housing department's checklist for the comparable stage lists the project manager's task as:

Brief
- review and sanction finalised brief.

Programme
- agree programme for design with consultant, including action and deadlines
- implement regular and frequent progress monitoring mechanism with consultant and quantity surveyor
- advise client of significant design changes and cost implications.

Consultations
- collect and collate responses (constraints, parameters, etc, from interested parties)
- interpret and pass to consultant
- instruct consultant regarding preliminary surveys, research, etc
- consult users regarding
  - scheme
  - programme as it affects them.

Quality/cost decisions
- determine design parameters
- quality standards
- life-cycle cost policy
- instruct consultant to prepare 'costed options'
- consider scheme design/s and 'cost options'
- approve preferred option
- obtain 'scheme and estimate' approval of preferred design
- sanction preparation of detailed design.

Scheme and detailed design information (stages D to F) Stages D to F are concerned with developing the client's requirements which were defined in stage C, to the extent that the design can be 'realised' by the builder.

During these stages the cost implications of design decisions must be regularly and frequently monitored to ensure that the cost ceiling is not exceeded as discussed in section 1 page 8.

Tasks to be undertaken during stages D and F include:
- determining the form of the planning application (which may include discussion with bodies such as the Fine Arts Commission, local amenity societies, etc, in addition to local authorities)
- obtaining consents from freeholders and lease holders (party wall agreements, compliance with restrictive covenants, etc)
- preparing scheme and detailed drawings
- checking that statutory undertakers provide required services and equipment
- checking design against site survey
- checking soil reports and extending tests
- obtaining building regulation and planning consents
- testing viability of alternative services engineering schemes
- consulting insurers, engineers and fire officers
- checking means of refuse disposal, cleaning, etc
- ensuring compliance of design with codes of practice, appropriate standards, etc
- preparing specifications, contract documentation, liquidated damages sums, performance bonds, etc
- monitoring design development against programme regularly and frequently to ensure that progress is maintained
- test project viability in context of current political and economic climate and modify (or abort) as appropriate
- select procurement path and determine appropriate contract documentation in conjunction with consultants.

The local authority housing department's checklist for the comparable stage sets out the project manager's tasks as:

Specification
- consider materials and components
- instruct consultant to confirm availability of above
- consider use of specialist contractors
- instruct consultant to confirm specialist contractors' capacity to undertake work within time-frame
- instruct consultant to inform if any materials and/or components require precontract commitment and to take steps to ensure timely delivery.

Client's instructions involving design changes
- advise consultant of changes in client's requirements
- obtain prompt advice from quantity surveyor regarding cost implications of design changes
- sanction incorporation of changes if within budget
- advise client of cost implications
- confirm finalised scheme with client.

Instruction to progress to construction stage
- sanction completed design
- sanction procurement arrangements
- instruct consultant to proceed with appointment of contractor.

When a traditional contractual arrangement is used, the contractor will not be involved in stages A to F.

1.2 Cost planning framework

The objectives of cost planning may be defined as:

1. Ensuring
   (a) the client receives value for money
   (b) the sensible expenditure of available monies between all parts of the building by relating the design and specification to the client's budget, ie to produce a balanced design solution.

2. To keep total expenditure within the amount agreed by the client.

   The principles of cost planning are that:
   (a) there is a standard reference point for each definable part of the building
   (b) the system enables performance characteristics to be related to each reference
   (c) the system allows cost to be apportioned in a balanced way throughout the building
   (d) it allows analyses of previous projects to be classified in a standard manner
   (e) it accords with design methods
   (f) it allows the costs to be checked as the design develops against the costs apportioned
   (g) it allows designers to take the necessary action before being too committed to any one design solution
   (h) it allows for design risk and price contingencies
   (i) it permits the costs to be presented in a logical and orderly way to the client from time to time as the design develops.

It is important that a balance is maintained between aesthetics and economics. The designer's creativity should not be inhibited, but designing within a budget requires considerable discipline on the part of the designer.[2]

Cost planning data An estimate, cost forecast or cost plan can not be more reliable than the data on which it is based.

At inception stage, the client is able to provide very little data on which the person preparing the forecast may base his estimate. Typically, only the building type and an approximate floor area is known. As the design develops, more and more data is available until at stage F the design is fully developed.

The estimate becomes increasingly more accurate as stage succeeds stage.

Section 1 of figure 1.4 indicates the level of accuracy at each of the design stages. The percentages shown should not be regarded as more than indicative. Section 1 also indicates the nature of the advice which it should be possible to provide at each stage.

Units of cost Various 'units' of cost data are available for estimating purposes at different design stages. These are shown in section 2 of figure 1.4. All are based on historical cost data which are analysed and synthesised to provide the estimate for the proposed building.

COST PLANNING FRAMEWORK

RIBA PLAN OF WORK Stage in design:-	A Inception	B Feasibility	C Outline Proposal	D Scheme Design	E Detail Design
1 Achieveable accuracy of cost advice	*preliminary cost advice *indication of cost range ± 20%	*determination of cost ceiling ± 10%	*outline cost plan within client's cost ± 7½%	*refined cost plan ± 5%	*final cost cost check ± 2½%
2 Types of 'units' of cost data available for estimating purposes	(a) cost/m² of floor area (b) cost/m³ of building volume (c) cost of functional unit for schools, hospitals etc.		(a) group elemental cost	(a) approximate bill of quantities (b) tenders from specialist contractors (c) detailed elemental costs	(a) bills of quantities & Unit rates
3 Data sources	(a) builders' final accounts for units 2A/B(a) & (b) (b) clients' standard costs for 2A/B(c) (c) Elemental Cost Analyses from Building Cost Information Service, <u>Building</u> and <u>Architects Journal</u>, the magazines, and builders' costing feedback		(a) as for units 2A/B(a) and (b)	(a) builders' costing data for units 2D/E(a) & (b) (b) specialists' tenders for Units 2D/E(c)	

1.4 Cost planning – advice, costing units and data sources

The cost per square metre of floor area, the m² unit, is the one most used.

The gross floor area is calculated by multiplying 'a' × 'b' × number of floors as indicated in figure 1.5. Both 'a' and 'b' are taken from the inner faces of the external walls. No deductions are made for partitions, voids, etc. The area is, in effect, the extent of the shelter being provided for the client.

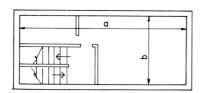

1.5 Calculating gross floor area

Approximate bills of quantities priced, using composite unit rates for measured work taken from *Spons* or similar price books, provide another method of cost estimating.

To make an approximate estimate of the cost of drainage, the drain runs might be measured as a single length by an average depth, say, 150 m at average 1.5 m deep. The rate would include breaking up and reinstating surface, excavation, drain bed, drain and surround.

The number of inspection chambers at an average depth might be priced using similar composite rate which would include excavation, base walls, cover, etc.

Bills of quantities prepared in accordance with a Standard Method of Measurement, priced with analysed and synthesised rates, provide a very accurate basis for cost estimating.

Sources of data A number of sources of data are available for estimating purposes. These are shown in section 3 of figure 1.4. The contractor's own costing data is one source.

A contractor may, for example, when reviewing his costs for a project at final account stage, divide the total cost by the gross floor area in order to calculate the cost in £/m² as the unit cost for the project being reviewed. The rate for the unit must be adjusted to allow for inflation, design, decisions and other factors if it is to be used for estimating future costs.

Some clients have 'standard' costs of 'functional unit' which they have derived from earlier projects and which they use to provide standards for future projects. For example, hospital boards set cost ceilings based on the 'cost per bed' for hospitals, education authorities on the 'cost per place' for schools.

The most detailed cost data for cost planning purposes is available from the BCIS and the magazines *Building* and *The Architects' Journal*.

Tenders from suppliers and specialist contractors are the most reliable data source for the estimator because they represent the actual price for which the tenderer is prepared to carry out specified work.

Cost planning practice – inception to detailed design (stages A to F)
During inception and feasibility stages (stages A and B), when only the floor area and building type are known, cost prediction must necessarily be crude. The units of available cost data are shown in section 2 of figure 1.4. The level of accuracy becomes greater as the client develops the brief. As shown in section 1, the level of accuracy should be ±10% as stage B is concluded.

To calculate the m² cost, the total cost of the building is divided by the gross floor area, eg to calculate the cost/m² for the building shown in figure 1.27%

$$\frac{\text{total cost excluding contingencies}}{\text{gross floor area of building}} = \frac{£613874}{397\,\text{m}^2}$$
$$= £1546.26/\text{m}^2.$$

A more detailed and precise version of the 'm²' unit is based on the cost of the 'elements' of the building. In order to use this approach, the historic cost data must be analysed element by element. BCIS has designed a standard form of cost analysis which facilitates the presentation of cost data in elemental form. Figure 1.28 provides an illustration of such an analysis.

The 'cube', the cost/m³, of building volume is an alternative to the 'super'. To obtain the cost of the cube, the total cost of building is divided by the number of cubic metres in the volume of the building. The volume is calculated as shown in figure 1.6. The 'cube' method is not much used.

During the outline proposal stage (stage C), the client's requirements are determined and group element costs in cost analyses can be used in addition to the units of cost shown in section 2A/B. The level of accuracy increases from the ±10% at stage B to ±7½% in stage C.

During scheme and detailed design stages (stages D to F), more design data is avail-

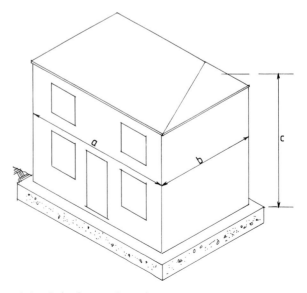

1.6 Calculating the volume

able. The cost estimating methods available increase as indicated in section 2D/E. Specialist contractors' tenders provide the most accurate basis for cost estimating for the reason given above.

1.3 Estimating and tendering

The CIOB *Code of Estimating Practice* defines estimating and tendering: *Estimating*, the technical process of predicting costs of construction, and *Tendering*, a separate and subsequent commercial function based on the estimate.

When the traditional contractual arrangement is used, the tender is the entry point for the contractor into the building procurement process.

Invitation to tender An invitation from a prospective client to a contractor to submit a tender for work is frequently the result of the contractor's efforts to exploit the market. It may also rise from the recommendation of a former satisfied client. The contractor has cause to be particularly pleased when invitations are from the latter source.

The invitation may take the form of a letter or telephone call asking if the contractor would be interested in tendering for a particular job in the near future or the first indication may be the arrival of a parcel containing drawings, specifications or bills of quantities. Specialist contractors are occasionally invited to advise on the type of construction or installation which might be suitable for a project as a prelude to a more detailed invitation.

The first decision which faces the contractor is whether or not he wishes to tender. His decision will be influenced by his corporate plan and his available resources. Manpower and cash availability are usually the principal factors which affect his decision.

If the contractor is disposed to decline the invitation but he believes such an action may prejudice his chances of being invited to tender on future occasions, he may decide to accept the invitation but take steps to ensure that his tender will not be accepted.

This may be done by submitting an inflated tender. The risk that the contractor takes in this event is that his tender will appear so unrealistically high that he will lose his credibility with the client.

Cover prices To avoid this risk, contractor 'A' may ask a competitor to suggest a sum which he, contractor 'A', may use as his tender sum. The competitor will obviously suggest a sum which is in excess of his own tender sum but contractor 'A' will be confident that the tender sum he will be submitting will appear to be competitive.

Giving and taking 'cover prices', as the practice described above is called, is not in itself illegal but it is discouraged in the building industry because it may lead to illegal collusion.

Opinions vary regarding the reactions of clients when their invitation to tender is declined. Most clients maintain that they would prefer contractors openly to decline an invitation and that such an action does not prejudice the contractor's future. Nevertheless, many contractors remain unconvinced in this respect and cover arrangements are not unusual.

Decision to tender The decision to tender is a decision to incur cost because estimating requires knowledge, experience, skill and time. With experience a contractor can make a shrewd 'guesstimate' of the cost of a job without numerous calculations and, if the profit 'margin' he can allow himself in his estimate is large enough, he can make a few errors on the 'swings and roundabouts' principle without coming to any harm.

If market conditions are favourable and the contractor is in a position to include a profit of, say, 20% then ±5% is neither here nor there. If, however, his profit margin is only 10% an error of 5% is much more significant and could make a difference to his expansion plans for next year. The contractor should, therefore, take care with his estimate. It is the foundation on which his success rests. Obviously, no contractor can afford to spend too much time on a tender because the odds against him getting it are, perhaps, 5:1, but he should remember that, if his tender is accepted, it makes sense if his estimate can be converted into his production plan.

A methodical approach to estimating The estimate may be of importance to the contractor not just during but after the works have been completed and the final account is being prepared. An estimator can only take into account the information which is available to the contractor at the time of tender. If the works are varied during the contract period from those for which the contractor tendered he is entitled to have the contract sum varied. Evidence of the information which was available for him at the time of tender may be invaluable when agreeing the cost of variations or claims.

The contractor should, therefore, take care when preparing his estimate to record all the documents used in its preparation, so that his estimate will be intelligible, perhaps, several years later. Care should also be taken to ensure that the estimate details are not mislaid during the course of the project. Much may hang on the contractor being able to prove that this or the other piece of information was or was not known at the time of tender.

If, then, the contractor decides to submit a tender he should plan his programme for the preparation of his estimate and the submission of his tender with care. A methodical approach is an essential ingredient of estimating. This section is concerned primarily with principles and method. The principles of tendering are common to all types of work which the contractor may be called upon to undertake.

Data for estimating Cost estimating involves the collection, collation, analysis and synthesis of data obtained from a variety of sources. Figure 1.7 shows the sources and the flow of data prior to preparing a tender.[3]

At the centre of figure 1.7 are the pretender procedures, shown in circles, 1.1 to 1.6, in figure 1.8. It is these procedures which are codified in the CIOB Code.

Overhead costs Overhead costs are those costs which are not attributed to the cost of production but which are, nevertheless, part of the cost of running the business. They are often separated into 'direct' and 'indirect' overhead costs.

Direct overhead costs These are costs which relate to a particular project. They include items such as site management and supervision, site accommodation and temporary services and plant of the supportive type referred to above. Direct overhead costs may be calculated on a weekly or monthly basis which may be converted to the total cost for the contract by relating the weekly/monthly cost to the estimated duration of the contract. The total cost of the direct overheads may be included in the tender as a lump sum or as a percentage addition to the unit rates in the manner suggested when referring to supportive plant costs above.

Indirect overhead costs Such costs are those which relate to the contractor's 'establishment'. These costs include his office rent, rates and the pay of his permanent staff such as secretarial support, payroll and accountancy staff, telephone operator, etc – the staff who will have to be paid regardless of the firm's level of activity. Overhead costs also include the cost of financing the firm. If the firm is running on an overdraft or loan, the interest payable on it should be paid, in part, by each project. Alternatively, if the firm is financed by the proprietor's or partners' own funds they are entitled to reimbursement at, at least, the rate of interest they could expect by investing their money elsewhere. Financing costs include the cost of financing the project itself.

Indirect overhead costs may be estimated for the forthcoming year/s on the basis of previous year's costs.

Each project should contribute a share towards the indirect overhead costs. Without the support of the firm's establishment there would be no contract! The costs are usually allocated to individual contracts pro rata the value of the firm's turnover. If, for example:

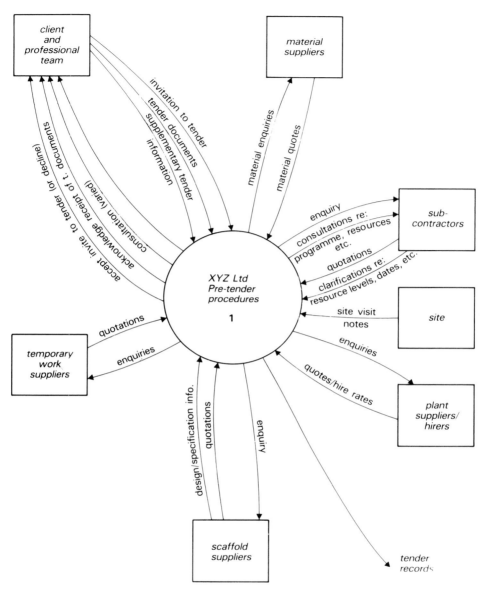

1.7 Data context for pre-tender procedures
FISHER, N, 1986

estimated turnover for
  the year is   £1,000,000
estimated prime cost of a
  contract is   £50,000
share of overhead cost for
  the contract would be

$$\frac{50,000}{1,000,000} \times 100 = 5\%.$$

If the estimated overhead cost
  for the year is   £90,000
the share for the contract
  would be 5% × £90,000 =   £4,500.

Alternatively, and perhaps preferably, the overhead cost may be expressed as a percentage of the turnover and used to cal-

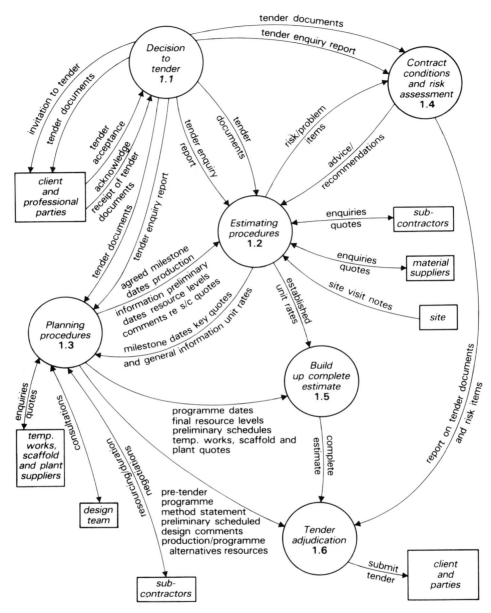

1.8 Pre-tender data flow
FISHER, N, 1986

culate the sum to be allocated to the contract:

$$\frac{£90{,}000 \text{ estimated overhead cost}}{£1{,}000{,}000 \text{ estimated turnover}} \times 100$$
$$= 9\% \times £50{,}000$$
$$= £4{,}500.$$

A contract's share of indirect overhead costs should be built into the tender so that it may readily be identified should the contractor's tender be successful.

The reason for requiring clear identification is that the estimate should be the budget sum for an ongoing project; clear

ESTIMATING AND TENDERING

identification of the cost components makes it simpler for the actual production costs to be monitored against the budgets to which reference is made later in this section.

Although the methods used to calculate overhead costs which are described above are quite widely used, they are really only directly applicable if the firm's turnover and overhead costs are unlikely to change greatly from one year to the next. Economies of scale will be achieved if overhead costs are 'pegged' while, at the same time, turnover is increased. Conversely, a reduced turnover without a corresponding reduction in overhead costs will make the firm less competitive in simple estimating terms.

There is, then, some flexibility of approach to overhead costs and a contractor keen to obtain work may be prepared to tender for work with a smaller addition to his production (prime) cost for overhead costs than the calculations outlined above would indicate as being appropriate. He cannot, however, do so lightly or for a prolonged period without inviting insolvency.

Profit margin – the commercial decision
Theoretically, numerous factors affect the margin such as demand for the product, competition, etc. In practice only a few firms attempt a scientific analysis of the profit margin which they hope should secure the project – just.

Analysis of competitors' profit margins The cartoon in figure 1.9 illustrates the 'intuitive' approach to profit margin calculations. Experience has shown that time spent by the estimator in recording his firm's (company A's) tender data and comparing it with the tender sums of company A's competitors enables him to make better informed predictions of profit margins which may secure future contracts. Most

1.9 The intuitive approach to profit margin calculation

contractors are aware that all too often the successful competitor submits a tender which appears to be uneconomically low; occasionally less than the figure which the estimator believes to be his firm's prime cost. No amount of comparative analysis of tenders will assist the competitor to compete with 'idiot' tenders; the contractor must hope that their folly will remove them from his list of competitors. But analysis of the 'sensible' competition frequently indicates that company A could increase its profit margin by several per cent and continue to submit competitive tenders. Analysis of competitors' profit margins makes it possible for the contractor to obtain some indication of 'what the market will stand'.

One approach to analysing tendering efficiency is put forward by Park.[4]

The basis of his approach is that in a competitive tendering situation the highest price that a contractor can set for his work is the lowest price that his cheapest competitor is willing to take to do the job.

If this is so, tendering efficiency may be defined as the ratio of the amount of profit actually made to the amount that could have been made had all the competitors' tenders been known in advance. Alternatively, it represents the amount which could have been made by taking all the jobs at the lowest competitor's price, provided the jobs would have been wanted at those prices. The concluding proviso is significant and should not be ignored.

To adopt Park's approach the contractor should keep a record of his tenders with the essential data tabulated under the headings shown in figure 1.10. Only four of the nine projects for which the contractor has kept records have been included in figure 1.10 so the totals of the figures in the estimated cost, potential and actual profit columns will appear not to tally.

Assuming nine recent projects, on four of which the contractor submitted the lowest tender, estimated direct costs total is £302,800 and a profit of £26,700 assuming the cost estimates were correct. His additions for profit on the nine projects varied from 5 to 14%. The lowest competitor's tender for each project determines the maximum profit that can be made on that job so that the 'maximum profit potential' represents the difference between the lowest competitor's tender and the estimated project cost.

Therefore, the maximum profit potential defines the most profit that could possibly have been made. Even if all the competitors' tenders had been known in advance for these nine projects, the maximum profit potential was £107,100. The £26,700 profit actually realised by the contractor on the four projects he carried out represents 24.9% of his maximum profit potential on the nine projects:

$$\frac{£26,700}{£107,000} \times 100 = 24.9\%.$$

A tendering efficiency within the range of 20 to 30% is typical of many contractors who operate intuitively.

The table in figure 1.11 shows the effect of the contract applying different additions for profit.

Between the extremes at 0 and 20%

Project number	Number of competitors	Lowest competitor's tender (£)	Estimated cost (£)	Tender sum (£)	Maximum profit potential (£)	Actual profit (£)
1	2	95,400	87,600	96,300	7,800	0
2	5	41,400	36,300	40,600	5,100	4,300
3	4	416,500	388,300	428,900	23,200	0
and so to						
9	5	478,900	457,500	498,500	21,400	0
			1,884,000		107,100	26,700

1.10 Record of tenders

ESTIMATING AND TENDERING

Percentage addition applied to project	Number of projects won	Estimated cost of projects won (£)	Gross profit on projects won (£)	Tendering efficiency (%)
0	9	1,884,000	0	0
1	9	1,884,000	18,800	17.5
2	8	1,250,900	25,000	23.3
3	8	1,250,900	37,500	35.0
4	8	1,250,900	50,100	46.7
and so to				
12	3	170,900	20,600	19.2
15	1	22,400	3,400	3.2
20	0	0	0	0

1.11 Tendering efficiency calculation

addition for profit, different addition will produce, for example, £37,500 at 3% and £20,600 at 12%. A low percentage addition will produce more projects without, necessarily, a corresponding increase in profit, as the table shows. In the example given above, a 7% addition would win six-ninths projects and produce a profit of £54,500 but an additional 1% addition would reduce the profit to £31,300. At 7% addition the contractor would have a tendering efficiency of 50.9%. There is clear scope for experimentation and a larger sample of projects would produced more statistically significant results.

Break-even calculations This method of calculating profit margins is more appropriate for production-line manufacturing than for 'unique' product manufacturing of the type undertaken by many contractors. Nonetheless, some specialist contractors exist, primarily, because they act as an adjunct to a production line so the method is not inappropriate.

For purposes of break-even method, costs are considered under the headings of overhead costs and unit costs.

Overhead costs include the items discussed earlier in the section. For a manufacturer, overhead costs would include the cost of running the factory and the office costs.

Unit costs would include labour, materials, fuel and costs for each unit produced.

The overhead costs plus the unit costs comprise the expenditure cost which may be measured against the income or revenue from the sale of the products to ascertain the point at which the expenditure and revenue break even.

The mathematical solution is not difficult but a graphical solution makes it easier to demonstrate trends on which to base management decisions.

The graph shown in figure 1.12 assumes that the overhead cost for the period is £90,000 which is shown as a horizontal line. The period used for the graph may be the day, week, month or year. In figure 1.12 the cost for the year has been used. On the horizontal axis is plotted the planned output for a contractor who manufactures windows as part of the group's activity.

The unit cost line has its point of origin where the vertical axis meets the overhead cost line. In the graph it is derived from an assumed unit cost of £208.33 per window. The revenue line is plotted using the sale price of each unit of sale. The sale price is determined by market conditions. In this example the sale price is assumed to be £283.00. In the graph, revenue will break even with expenditure when 1,201 units have been sold and from that point forward the contractor will make a profit. An advantage of the graphical presentation is that management, knowing the number of units which must be produced before a profit will be made, is able to plan its sales and calculate profit margins on the basis of costs. For products which have a relatively low unit cost and a relatively high overhead

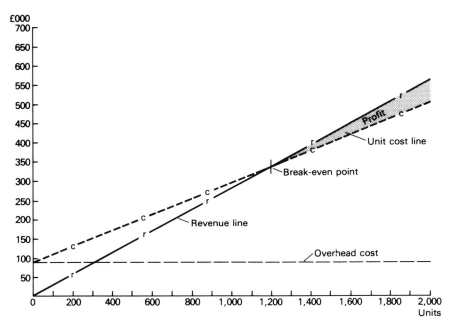

1.12 Break-even analysis

and development cost – books, for example – the unit sale price may be greatly reduced when sufficient copies have been sold – hence the 'remainders' on offer in railway station shops and street markets.

Submission of tender When the estimator has estimated the cost of the labour, plant and materials to be used and ascertained the duration of the work, the management team, proprietor or board of directors (as the case may be) will normally become involved in the decision-making process prior to submission of the tender.

The CIOB estimating and tendering practice The CIOB Code of Estimating Practice provides an authoritative guide to good practice in estimating for building work from preselection to the acceptance of a successful tender when traditional contractual arrangements are adopted.

It distinguishes between *estimating* and *tendering* and considers the formal steps to be taken in converting the net cost estimate into a tender at the adjudication stage.

The basic principles of estimating are set out in the code, which considers the integration of estimating and management and links with the functions of programming, buying and construction.

Definitions and outlines of the stages, taken from the code, are given below:

Preselection is concerned with the establishment of a list of contractors with suitable experience, resources, ability and desire to execute a proposed project, bearing in mind the character, size, location and timing of the project.

Final selection is usually made either by negotiation or tender on the whole or part of the works, once the list of contractors has been determined by preselection procedures.

Management is an important aspect in the production of the estimate and the subsequent tender. The estimator must have management responsibility within the estimating department and a responsibility for managing the production of the estimate. This responsibility requires the estimator to undertake various actions in managing resources and information during the tendering period.

This management role must not be underestimated. The estimator has to ensure that other company departments and staff work to his requirements, produce information on time and in the format required, and that effective operating procedures and lines of communication are established between departments to allow the efficient production of the estimate.

The code recognises that firms have their own procedures and preferences but it identifies the following stages in the estimating and tendering processes:

- preselection
- decision to tender
- project appreciation
- making enquiries and obtaining quotations
- calculating all-in and unit rates
- completing the cost estimate
- estimator's report and adjudication
- action after submission of a tender.

The decision to tender can occur at one of two stages:

- when preselection enquiries are initiated by a client or his consultants, the contractor will make a decision based upon an outline of the tender information available at that stage. This intention to submit a tender must be re-affirmed when the full invitation to tender and supporting documentation are received
- when the preselection procedure has not been followed, the contractor may find that tenders arrive without prior notice. In such instances, only one opportunity exists to appraise the project and make the necessary decision to tender for a project or not.

In the case of a project where preselection has occurred and details have already been sent to the contractor, a checking procedure is needed to confirm that the project conforms with the information already provided and that the contractor's position regarding tendering has not changed. It is essential that adequate time is allowed for this procedure by clients and consultants. In either situation, the contractor will follow the same procedures in analysing the information received.

The flow chart in figure 1.13 outlines the activities involved in making the decision to tender.

Project appreciation commences in the 'decision to tender' phase but intensifies once management has confirmed that an estimate is to be prepared.

Considerable management skills and personal leadership are needed by the estimator to motivate and co-ordinate the various staff associated with the preparation of the estimate and its subsequent conversion into a tender.

Co-ordination meetings may well be necessary with management and other departments within the contractor's organisation to establish key dates, decide on actions necessary and monitor progress during the production of the cost estimate.

The flow chart in figure 1.14 outlines the activities involved in project appreciation.

Enquiries and quotations The contractor's success in obtaining a contract can depend upon the quality of the quotations received for materials, plant and items to be sub-contracted. It is essential to obtain realistically competitive prices at the time of preparing the estimate.

The contractor must ensure that comprehensive records are maintained of the various elements of project information sent to suppliers and sub-contractors. These records must list the drawings sent, the relevant contract and specification clauses, project preambles and the pages of bills of quantity.

The use of standardised documentation and procedures will assist in making this stage in preparation of the estimate routine, and allow an interchange of personnel at any stage.

Calculating all-in rates and unit rates is an important part of the estimating process which is described in detail in the code but is outside the scope of this text.

Completing the cost estimate involves resolving information which may have been unclear, making adjustments for compe-

20 CONTRACT PRINCIPLES, PROCEDURES AND PRACTICE

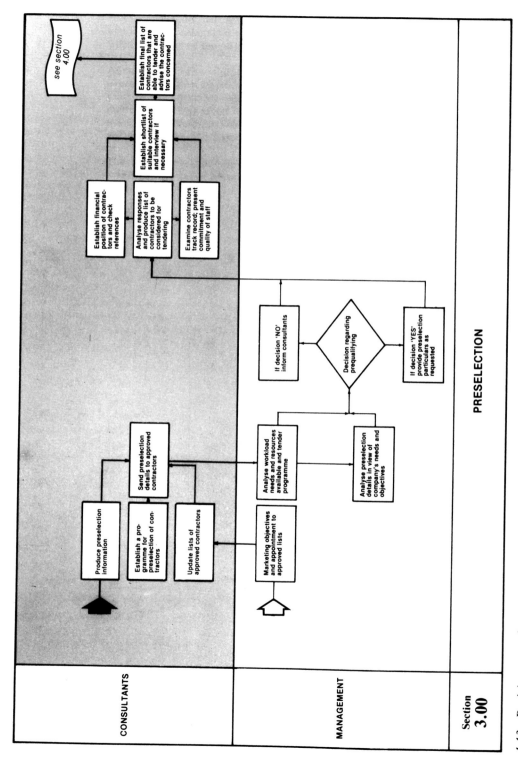

1.13 *Decision to tender – activities*
CIOB code

ESTIMATING AND TENDERING

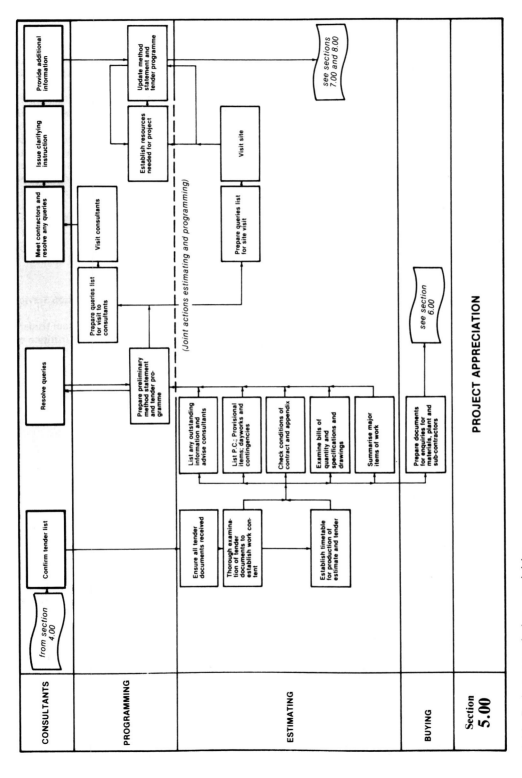

1.14 *Project appreciation – activities*
CIOB code

22 CONTRACT PRINCIPLES, PROCEDURES AND PRACTICE

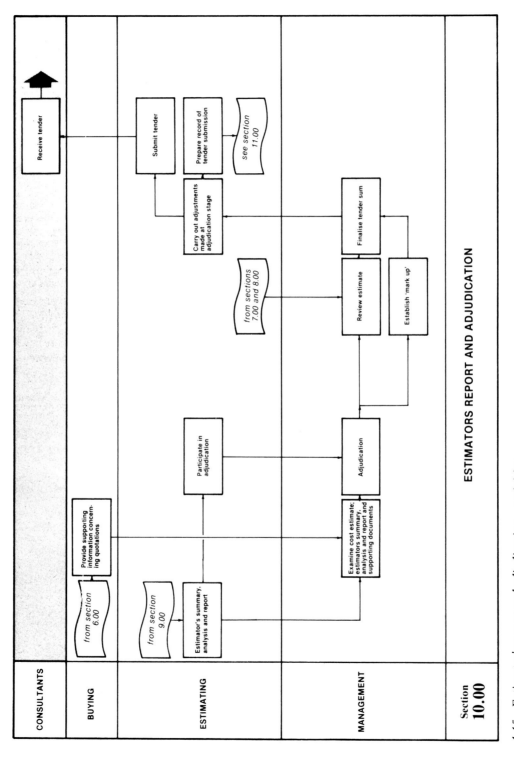

1.15 *Estimator's report and adjudication – activities*
CIOB code

titive quotations which arrived after the estimate had been summarised, and estimating the cost of project overheads, preliminaries and conditions of contract.

Provision must be made for a fluctuating or firm price tender.

Provision must also be made for finance charges during the course of the works.

The 'technical estimate' stage concludes with the estimator's own review of the estimate and finalisation of the cost estimate prior to adjudication.

Estimator's report and adjudication The estimator's summary analysis and report brings together all pertinent facts which have influenced the preparation of the estimate for adjudication by management. The objective of this report is to highlight to management the various matters which have been identified as cost significant, where alterations have been made to normal production standards and any special or unusual contract conditions or risks.

The flow chart in figure 1.15 outlines the activities involved in completing the estimate, its conversion to a tender and submission of the tender.

Action after submission of tender is concerned with assessing tenders and notifying results, adjustment of errors, and action to be taken with a successful or with an unsuccessful tender.

In order to facilitate analysis of competitors' profit margins in relation to his own firm's, discussed above, the estimator should ascertain and record competitors' tender sums.

Cost data cycle The contractor's estimate should be prepared with a view to it providing the basis for a successful tender but, also, with a view to the future.

It provides the basis for planning, monitoring and controlling production cost and progress, for calculating productivity-based incentive payments and, if necessary, for the preparation of applications for pay-

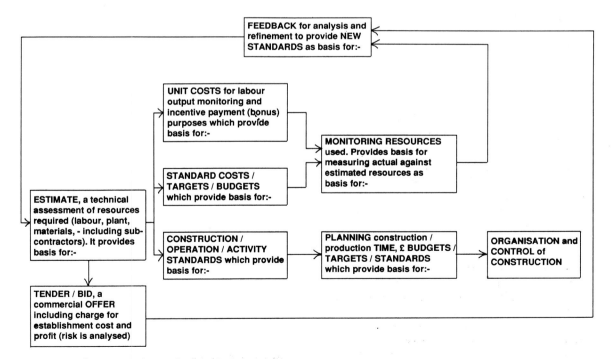

1.16 Production and cost data cycle for construction projects

ment of loss and expense should the regular progress of the works be disturbed.

The format of the estimate should be clear and detailed and all the supporting documentation should be retained. Should there be a dispute regarding cost the contractor may be required to produce his estimate in order to prove his loss.

Figure 1.16 outlines the cycle followed by production and cost data.

At the estimate stage, shown on the left side of the figure, the estimator uses data fed back from previous projects and other sources to make a technical assessment of the resources required to build. The resources comprise labour, plant, materials, sub-contractors, etc. The quantity and type of resources and estimated costs provide the *standards* used to plan, monitor and control the works during the construction stage. Further references are made to the production and cost data cycle later in the text. The cycle is fundamental to the construction process.

Section 1B
MOBILISATION OF THE WORKS

1.4 Sub-contractor selection

Precontract negotiations Reference is made below, when discussing nomination procedures, to the importance of all parties involved having similar expectations of conditions, terms and the nature of the attendance which will be provided to the sub-contractor, etc, before the contractor and sub-contractor enter into contract. The importance of similar expectations applies as much to domestic as to nominated sub-contracts and the headings in NSC/1 provide a useful guide when negotiating with domestic sub-contractors.

It is invariably easier to reach agreement before entering a contract than after. Attitudes harden remarkably quickly once the contract has been signed.

Selection The contractor's performance depends on the performance of his sub-contractors (and suppliers) so the greatest care should be taken with the selection of sub-contractors. If the contractor is not satisfied that a sub-contractor whom the architect proposes to nominate (or name) will be able to meet the contractor's performance requirements, he should withhold his agreement to the nomination. The contractor should not withhold such agreement 'unreasonably' (JCT 80 cl 35.1.4) but well documented and presented reasons could not be regarded as being unreasonable behaviour on the part of the contractor.

When selecting sub-contractors, the contractor should look for in the sub-contractor the strengths and weaknesses which he would look for in himself when carrying out a corporate review.

Sub-contractors should, then, be selected having in mind the three requirements for client satisfaction, namely, that the client receives a building which is fit for his purpose (complies with his specification), that it is within his predetermined cost limit and that it is delivered on time.

Clearly, the sub-contractor's competitive price is important but it is only one (and not necessarily the most important) requirement. The installation, system or whatever must also perform satisfactorily and be delivered on time.

Factors affecting selection A list of factors which may affect the selection of a sub-contractor might include:

Financial status Sources of information comprise:

- *Bankers' references* If such references are available they provide a quick source but they may not provide sufficient detail for the contractor's purposes.
- *The register of companies* The contractor may obtain annual reports of private and public limited companies. From these it is possible to make detailed analyses of a company's financial strengths and weaknesses but companies do not always submit their accounts promptly and even if they are submitted within the

statutory period they may not be sufficiently up to date to provide the contractor with a reliable guide to the sub-contractor's present financial condition. In an industry as volatile as the construction industry, up-to-date information is important.
- *Credit agencies* specialise in providing intelligence on the financial status of businesses for a fee. Their intelligence is up to date and detailed analyses will be provided if required. Their services are provided quickly and they are extensively used. They provide ratings which indicate credit limits appropriate for the firm being assessed.

Whatever the source of the data, criteria must be provided for assessment.

Financial assessment of a firm may be made using Standard Accounting Practices which provide a number of traditional ratios to ascertain performance, profitability, liquidity, share value, etc. The ratios are those used generally for inter-firm comparisons.

Detecting potential corporate failure Accounting Practice 1 provides a subjective method of looking at the non-financial signs of corporate failure. It requires observation, meeting the directors and forming views. The method assumes that failure is a progressive condition over a period of years, that managerial errors occur and that there are discernable defects in management attitudes and accounting systems.

Defects These occur under three headings: managerial, accounting systems and change.

Managerial defects are largely concerned with autocracy on the part of the chairman, managing director or chief executive and a passive or non-contributing board of directors who lack managerial or other skills. There is a lack of financial direction in middle and lower management skills.

Accounting system defects are generally concerned with lack of budgetary planning, monitoring and control. There is an absence of regular short-term aims and cash flow management. The absence of a costing system makes it impossible for the manager to know unit costs.

Failure to recognise and adapt to change to meet clients' requirements indicates a disregard of market trends. Such sub-contractor's will probably lack modern equipment and methods.

Mistakes which may lead to the sub-contractor's failure include disregard of weaknesses indicated by financial performance ratios, overtrading and undertaking projects which are beyond its capability.

Accounting Practice 1 provides a weighted scale which enables the observer to assess strengths and weaknesses.

Previous experience Selection based on the contractor's previous experience of the sub-contractor, if any, may provide useful information regarding the sub-contractor's future performance.

A typical list of factors or attitudes on which the contractor's staff comments may be sought are:

Factor or attitude	Weight
Co-operation and service	6
Managerial experience of the sub-contractor's staff	10
The quality of its site supervision	6
Technical expertise	10
Quality of materials used	8
Workmanship quality	10
Industrial relations	6
Quality of safety procedures	10
Ability to commence and complete to time	10

The importance that the contractor places on the above factors and/or attributes will determine the weighting given to each. Indicative weightings are suggested.

The experience of contract and site managers, surveyors and others may be sought and tabled.

Assessment procedures should be developed to include:

- visit to the sub-contractor's office and works to observe their equipment and

- mechanisation and to obtain a general impression of competence, condition of fabric, etc
- an assessment of the defects and mistakes referred to above
- an analysis of the financial ratios to determine status
- comparison of sub-contractor's tenders
- consideration of sub-contractor's factors and attitudes, listed above
- interviews with a shortlist of potential sub-contractors to make an assessment of their staffs' attitudes and abilities to work as a team with the designers, main contractor and other sub-contractors.

1.5 Entering into contract

Entering into contract marks a positive commitment by the parties to the contract.

The initiative rests with the client. The tasks to be undertaken by the project manager indicate the activities involved at this crucial stage.

Project management tasks undertaken when entering into contract include:

- agreeing with the client that a contract is to proceed to the construction stage
- ascertaining that the site is available so that the contractor may be given possession
- determining the procurement path and ensuring compliance with EEC requirements
- making arrangements regarding preparation of contract documentation, legal requirements, etc
- determining the tender period and timetable and sanctioning tender or other procedures, agreeing names of appropriate firms and personnel and ascertaining their willingness and availability
- agreeing date for possession
- ensuring consultants and contractors know full extent limits of the site to be possessed
- supervising tendering action, considering contractor's enquiries, receiving tenders and entering into contracts
- advising unsuccessful tenderers of tender results
- arranging pre-contract meetings with consultants, tenants and adjoining owners, etc.

The contract provides a framework within which the parties must operate. The law relating to contract is beyond the scope of this text but the following notes may act as a reminder of the more significant aspects of the contractual framework.

Aspects of contract A contract is an agreement which is enforceable by law. A contract contemplates and creates an obligation. The essentials in the formation of a contract are:

- agreement between the parties
- an offer
- an acceptance of the offer
- a consideration (the price to be paid for the service or goods to be provided).

An agreement between the parties regarding the purpose, rights and the obligations which the contract will create is essential. In order to create an obligation the agreement must:

- be for a lawful purpose
- be between parties who have the capacity to enter into a contract
- be free from misrepresentation, mistake, duress or other circumstances which will vitiate (invalidate) it.

A contract is effected when an offer made by the offerer is accepted by the offeree. An offer may be in writing or it may be oral. It may be in the form of a gesture as occurs during an auction when the bidder nods to the auctioneer to make a bid.

An essential aspect of an offer is that the offerer intends to be bound by the terms if the offer is accepted. In this respect an offer is different from a declaration of intention and a statement of price which is known as an 'invitation to treat'.

An invitation from a prospective client to a contractor to 'give him a price' (to tender for work) is an example of an invitation to treat. If the contractor offers to carry out the work for £x and the client accepts the contractor's offer, an agreement has been made and a contract has been created. Both, or either, offer and acceptance may be oral or in writing – the contract will be binding on the parties concerned.

A difficulty which arises when the parties wish to enforce an oral contract is establishing the parties' intentions. To Sam Goldwyn, Hollywood film magnate, is attributed: 'an oral contract is not worth the paper it's written on'.

A simple exchange of letters, the first making the offer, the second accepting it, will prevent future misunderstandings.

Figure 1.17 illustrates the contractual relationship between the parties engaged on a typical building project.

Express and implied terms Express terms are those statements or promises which are made by the parties to the contract. An undertaking by the contractor that he will complete the works by a certain date, or by the client that he will make payments on account when certain stages of construction are reached, are examples of express terms.

Implied terms may arise by custom or trade practice and by statute or legal precedents established in the courts from cases which have been contested in the past. An 'implication' is that goods or services to be provided will be fit for their purpose.

An express term will normally take precedence over an implied term if, for example, the express term contradicts a

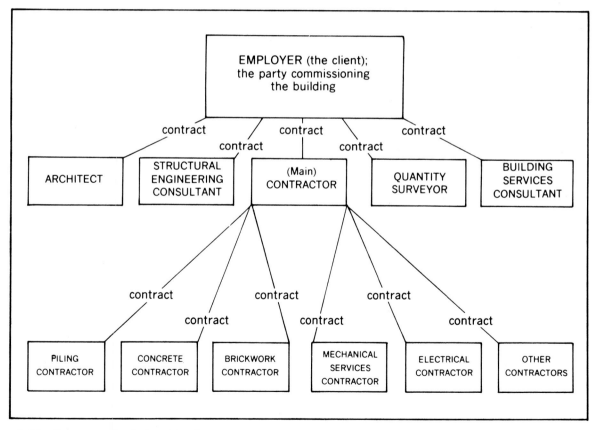

1.17 Contractual relationships between parties

trade custom but an express term cannot override a legal obligation.

Privity of contract Only the parties to a contract may normally benefit from or be bound by the terms of a contract. A third party cannot normally sue upon a contract even if the contract has been made expressly for the third party's benefit.

Assignment Parties to a contract may not assign their benefits or obligations to a third party without the consent of the party with whom they have entered into a contract. Conditions in a standard form of contract usually require agreement between the contracting parties as a condition of assignment.

Standard forms of contract have the advantage that with time their conditions become familiar to the parties so that they 'know where they stand'. If the parties intend to adopt a standard form of contract they are well advised to do so before commencing the works. The parties, particularly the contractor, may be vulnerable to pressures once the works have commenced. An oral contract or exchange of letters will not, for example, ensure his entitlement to payments on account during the progress of the work.

The parties should be aware of the conditions before entering into contract. The rule is to *think (and know) before signing and sign before starting*.

Reasonable is an adjective which appears frequently in the conditions of contract. What is to be regarded as a reasonable standard, time, etc, is almost invariably a matter for the opinion of the architect. Should his opinion as to reasonableness be disputed by the contractor it would be necessary to refer to a third party. Section 4 of this text discusses the resolution of disputes.

JCT 80 has been selected for the purposes of the following text because it is the form of contract most generally used and because it is the most comprehensive. Other forms might be used. The aim of this section is to indicate the nature of the conditions contained in standard forms of contract.

The review of the conditions of contract is intended as an introduction to contract procedures and administration. Many clauses have been paraphrased and to some extent simplified. In practice, the relevant clauses should always be read in full if the contractor has any doubt regarding interpretation or meaning.

The review is concerned with the more significant clauses. Those which the practioner or contractor is less likely to encounter are given only passing comment.

JCT 80 The standard form of contract comprises:

– articles of agreement
– conditions
– appendix.

The articles of agreement provide a proforma in which should be inserted the names and addresses of the employer and contractor, and the title of the 'Works' which the employer desires.

There are four *recitals* in the first of which it states that the employer 'has caused Drawings and Bills of Quantities showing and describing the work to be done to be prepared by or under the direction of' a person whose name is inserted. This person would normally be the architect.

The second recital states that 'the Contractor has supplied the Employer with a fully priced copy of the said Bills of Quantities' which are to be referred to in the conditions as the *Contract Bills*.

Space provision is made in the third recital for insertion of the numbers of the drawings which are to be referred to as the *Contract Drawings*.

The words 'Contract Bills' and 'Contract Drawings' and other terms have initial capital letters. They should be signed by or on behalf of the parties. The parties should take care to ensure that any revision letters of the numbered drawings are entered in the third recital.

The fourth recital makes reference to the need for the status of the employer for the purposes of the statutory tax deduction scheme under the Finance (No 2) Act, 1975, to be inserted in the appendix.

The second part of the articles of agreement comprises five articles which state:

- the contractor's obligations to be to 'carry out and complete the works shown upon, described by or referred to' (in the contract documents) (Article 1)
- the contract sum which the employer will pay to the contractor (Article 2)
- the meaning of the terms 'the Architect' and 'the Quantity Surveyor' in the conditions of contract (Articles 3 and 4) and that:

'if any dispute or difference as to the construction of the contract or any matter...shall arise between the Employer or the Architect on his behalf and the Contractor...it shall be and is hereby referred to arbitration in accordance with clause 41.' (Article 5)

The final page of the articles provides a proforma for the signatures and or seals of the employer, contractor and their respective witnesses.

Further reference to completion of the appendix is made below when considering individual clauses in the conditions of contract.

The appendix also provides a number of spaces against clause numbers in which the parties to the contract should insert dates, sums or forms of words to specify their intentions and or indicate which of a number of alternative clauses are to apply.

It is important that the items in the appendix are agreed and completed and that where alternatives exist the parties preferences are clearly stated.

1.6 Nominated, named and domestic sub-contractors

Alternative methods of selection Typically, sub-contractors may be nominated or named by the architect or appointed by the contractor with the agreement of the architect and contractor.

When JCT 80 is used, procedures regarding sub-letting are set out in clause 19. The alternatives are:

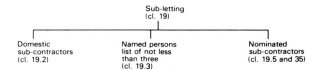

Domestic sub-contractors (with architect's consent) The main contractor 'shall not without the written consent of the architect sub-let any portion of the works' (cl 19.2) but subject to that proviso there is nothing to stop the main contractor employing any sub-contractor as a 'domestic' sub-contractor to carry out specialist works which are measured in the bills of quantities.

Named persons Under clause 19.2 the architect takes no part in the selection of the sub-contractor — the choice is the main contractor's provided the architect consents to the selected sub-contractor. Clause 19.3 states that where the contract bills provide, certain work measured or otherwise described in those bills and priced by the main contractor *must be carried out by persons named in a list* in, or annexed to, the contract bills and selected therefrom by, and at the sole discretion of, the main contractor. The 'persons' in the list are named by the architect and the list must comprise not less than three persons (cl 19.3.2.1). The architect keeps overall control of the firms (persons) who will act as sub-contractors but the main contractor is also able to exercise some control over the final selection because it is his 'say' which decides whether sub-contractor A, B or C is employed.

Nominated sub-contractors Reference to the nomination of sub-contractors appears in clause 19.5.1 of SF 80, which states that the provisions of the contract relating to nominated sub-contractors are set out in Part 2 of the conditions.

Part 2 of the conditions is concerned with both nominated sub-contractors and nominated suppliers. Reasons for nominating sub-contractors were outlined by the Banwell Committee in 1963 and still hold good (Banwell, 1964)[5]. Sub-contractors may be nominated when:

(a) the architect requires *special techniques* to be used; techniques which are the province of a specialist (sub-) contractor
(b) it is important to place an order for specialist work at an *early date*, probably before the main contractor has been selected
(c) a *particular quality* of work is required which is the province of a specialist.

Contractor's agreement to nominated and/or named sub-contractors Reference is made in section 1.4 to factors which may influence the contractor's selection of domestic sub-contractors. These factors include the proposed sub-contractor's financial status and his capacity to carry out and complete the sub-contract works. Such factors should also be considered before the contractor agrees to nomination or naming. He must always have in mind that it is he who has responsibility for the sub-contractor's performance.

Nominated procedures

Basic method There is a *basic* method and an *alternative* method of nomination (see page 39) which can be used by the architect. It is generally agreed that the *basic* method should be used for the vast majority of contracts and especially where major specialist works are concerned or where the sub-contractor makes a significant contribution to the design of the works. The 'operations' to be carried out are shown in figure 1.18 in network form. The numbers which appear below the line are the relevant sub-clauses in JCT 80. Follow through the operations to ascertain the sequence of events before considering the detailed implications.

Operation 1–2, 'architect selects sub-contractor'. Selection can be made before or after the main contractor has been appointed. Having selected the sub-contractor the next two operations run, largely, concurrently.

Operation 2–3 requires agreement from the sub-contractor to eventual nomination and involves completion of tender NSC/1 which consists of twelve pages. The aim of NSC/1 is to provide a means of negotiation between architect, sub-contractor and main contractor, so that when a contract is eventually entered into all parties have reached agreement on the terms.

Operation 2–4 requires the employer and sub-contractor to make an agreement so that the sub-contractor will be paid for design work, etc, should he subsequently *not* enter into a contract with the main contractor (see op 8–9), and so that the employer may use the sub-contractor's design if the sub-contractor is not employed to carry out the sub-contract works.

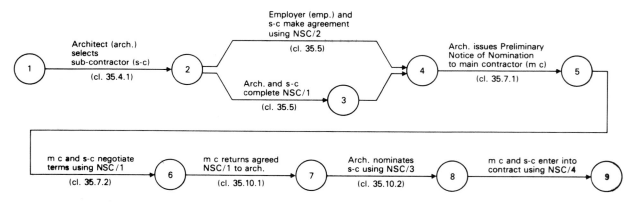

1.18 Procedure using Basic Method

Completion of operations 2–3 and 2–4 make it possible for the architect to issue a Preliminary Notice of Nomination to the main contractor (op 4–5) which is really an expression of the architect's intention to nominate the sub-contractor provided the sub-contractor and main contractor reach agreement as to details of attendance, contract periods, etc, which they attempt to do in operation 5–6.

Once terms have been agreed the main contractor returns NSC/1 (which will probably look rather dog eared by this time for reasons which will be gone into later) to the architect (op 6–7) which enables the architect to nominate the subcontractor using NSC/3 (op 7–8). The now nominated sub-contractor and main contractor enter into a contract using NSC/4 (op 8–9).

NSC/1, 2 and 3

Completing tender NSC/1 Having considered the sequence of events one may look in detail at the procedures using the 'operations' contained in the network (figure 1.18) as the basis.

Selecting the sub-contractor (op 1–2) The architect *may* use form NSC/1 for obtaining competitive tenders, or for negotiating with one sub-contractor, but its use is not compulsory. Any documents may be used for the *selection* of the sub-contractor. Once, however, selection has been made the architect and sub-contractor must complete NSC/1 which, together with sub-contract NSC/4, will eventually comprise the principal sub-contract document.

Completing NSC/1 (op 2–3) A complete specimen of page 1 of NSC/1 is reproduced as figure 1.19. From the printed wording on page 1 it can be seen that either a lump-sum quotation ('VAT – exclusive Sub-Contract Sum') or a unit rate quotation ('VAT-exclusive Tender Sum') with detailed quantities and prices may form the basis of the sub-contract. The parties should take care to ensure that any documents which have been used in calculating the sub-contract or the tender sum are clearly numbered and identified.

Page 1 also has space for daywork percentages to be entered. These percentages are to be added to the 'prime costs' which are defined by the negotiating bodies; the RICS, NFBTE, ECA, HVCA, etc.

When the architect and sub-contractor are satisfied that the 'sum' and percentages which have been entered on page 1 represent their intentions they sign at the foot of the page. It will be remembered that agreement NSC/2 is signed by the employer and sub-contractor at the same time as the architect and sub-contractor sign NSC/1 (op 2–4 in figure 1.19). Further reference is made to NSC/2 later.

The stipulations on page 2 of NSC/1 are important. The first explains the relationship of NSC/1 and NSC/2. It reads:

'Only when this tender is signed on Page 1 on behalf of the Employer as "approved" and the Employer has signed or sealed (as applicable) the Agreement NSC/2 do we agree to be bound by that Agreement as signed by or sealed by or on behalf of ourselves.'

The sub-contractor and architect may well sign on page 1 before the identity of the main contractor is known. Stipulation 2 reserves for them the right within 14 days of written notification by the employer of such identity to withdraw the tender and the agreement NSC/2 notwithstanding any approval of the tender by signature on page 1 on behalf of the employer. In stipulation 3 they reserve the right to withdraw the tender if they are unable to agree with the main contractor on the terms of schedule 2 of the tender.

The tender is withdrawn if the nomination instruction (nomination NSC/3) is not issued by the architect within an 'acceptance period' which is inserted in stipulation 4 *by the sub-contractor*.

The final stipulation on page 2 is that withdrawal under stipulations 2, 3 and 4 shall be at no charge to the employer except for any amounts that may be due under agreement NSC/2.

Pages 3–12 of NSC/1 comprise schedules 1 and 2.

Tender NSC/1

JCT JCT Standard Form of Nominated Sub-Contract Tender and Agreement

See "Notes on the Completion of Tender NSC/1" on page 2.

Main Contract Works: [a] Community Centre for Worsten District Council

Location: Worsten Causeway, Worsten, BEDS.

Job reference: CCW/29/82

Sub-Contract Works: Electrical Installation

To: The Employer and Main Contractor [a]

We J. Shine Electrical Services Ltd.

of 10 High Street,

Nearbigh, Beds Tel. No: 0123-45678

offer, *upon and subject to the stipulations overleaf*, to carry out and complete, as a Nominated Sub-Contractor and as part of the Main Contract Works referred to above, the Sub-Contract Works identified above in accordance with *the drawings/specifications/bills of quantities/schedule of rates for the Sub-Contract Works which are annexed hereto, numbered

Drawings Nos. CCW/1D, 2B, 3B, 4C, 7A Specification pp. 90-96 incl.

and signed by ourselves and by the Architect/Supervising Officer; and the Particular Conditions set out in Schedule 2 when agreed with the Main Contractor; and JCT Sub-Contract NSC/4 which incorporates the particulars of the Main Contract set out in Schedule 1.

*for the VAT-exclusive Sub-Contract Sum of £ 26,720.00

Twenty six thousand seven hundred and twenty pounds _____ (words)

*or on the VAT-exclusive Tender Sum of the ~~~~~~~~~~~~~~~~~~~~

~~~~~~ (words)

The daywork percentages (Sub-Contract NSC/4 clause 16·3·4 or clause 17·4·3) are:

Definition*[c]	Labour %	Materials %	Plant %
RICS/NFBTE			
RICS/ECA	100	15	15
RICS/ECA (Scotland)			
RICS/HVCA			

The Sub-Contract Sum/Tender Sum and percentages take into account the 2½% cash discount allowable to the Main Contractor under Sub-Contract NSC/4.

Signed by or on behalf of the Sub-Contractor   J Shine director for J Shine Electrical Services Ltd.   Date 14th August 1982

Approved by the Architect/Supervising Officer on behalf of the Employer   AD Signer RIBA   Date 22/VIII/82

ACCEPTED by or on behalf of the Main Contractor subject to a nomination instruction on Nomination NSC/3 under clause 35·10 of the Main Contract Conditions   A. Driver for P. Roper Builders Ltd.   Date 17 Sept. 82

1.19 Form of agreement NSC/1

The provisions of schedule 1 are designed to set out the main contract terms which are then part of the nominated sub-contract. Schedule 1 should reflect accurately the terms of the main contract executed between the employer and main contractor. This is particularly important where the architect and the proposed sub-contractor have completed, as far as possible, the tender section of NSC/1 before the main contract has been settled and executed.

The sub-contractor should pay close attention to the insertions which are made in schedule 1 because it is only the main contract terms which are notified in this schedule which are binding on him. There is reference to this in NSC/4, the form used by the main contractor and sub-contractor in operation 8–9 when they execute their contract. NSC/4 is the 1980 edition equivalent of the former Green Form conditions. There are two appendices to schedule 1 which are concerned with fluctuations and one or the other will be deleted depending on the method of adjusting variations of price which is to be used.

Schedule 2 of NSC/1 is titled 'Particular Conditions' and it is this schedule which provides the main contractor and the sub-contractor with a means to negotiate the conditions which will eventually be included in the contract between the two contractors.

In the words of the 'note' of schedule 2 the main contractor

'has to complete this schedule in agreement with the proposed sub-contractor. The completed schedule should then take account not only of the preliminary indications of the sub-contractor stated therein but also of any particular conditions or requirements of the contractor which he may wish to raise with the sub-contractor'.

Figures 1.20 and 1.21 show pages 9 and 10 of schedule 2 (it contains four pages in all) to illustrate how it is envisaged that they might be completed.

The architect's contribution is shown as typeface, that of the sub-contractor in longhand and the main contractor's insertions are in block letters.

Notes about which party is to complete which section are provided on page 2 of NSC/1. In figure 1.20 (page 9 of NSC/1) it can be seen that the period initially suggested by the architect in 1A as being when he envisaged the works will be carried out has been varied to some extent by the main contractor and electrical sub-contractor in 1C by mutual agreement. It may be that in the event contract progress will require the sub-contractor to bring forward or put back the periods when he will be able to work on the site. In 1B it can be seen that the parties have agreed that the sub-contractor will be given four weeks notice to commence on site.

In 3A (a) a compromise has been reached between main contractor and sub-contractor regarding the means of access to the 'high' area of the dining-hall. The initials of both parties opposite the acceptable wording can be taken as confirmation of their respective intentions. Clearly, all the parties to the contract will find it necessary to think through the project to a considerable extent at pre-contract stage, if they are to complete NSC/1 in a meaningful manner. The aim of pre-contract negotiation is to prevent, or at least substantially to reduce, post-contract misunderstandings.

Employer/Nominated Sub-Contractor Agreement NSC/2 The form 'Agreement NSC/2' is returned, completed, by the sub-contractor at the same time as his tender contained in NSC/1.

NSC/2 extends to four pages, two and a half of which are concerned with the *agreement clauses*. NSC/1 comes into operation at two stages; clauses 1 and 2 operate after tender NSC/1 has been approved on behalf of the employer by signature on page 1, and the remaining clauses after the *nomination* of the proposed sub-contractor on form 'Nomination NSC/3'. Reference to the implications of stipulation 1 on page 2 of NSC/1 in this connection was made earlier in this section.

The first part of NSC/2 is the *agreement* between the employer and the sub-contractor; the sub-contractor is *not a nominated* sub-contractor at this stage.

Schedule 2: Particular Conditions

Note: When the Contractor receives Tender NSC/1 together with the Architect/Supervising Officer's preliminary notice of nomination under clause 35·7·1 of the Main Contract Conditions then the Contractor has to settle and complete any of the particular conditions which remain to be completed in this Schedule in agreement with the proposed Sub-Contractor. The completed Schedule should take account not only of the preliminary indications of the Sub-Contractor stated therein, but also of any particular conditions or requirements of the Contractor which he may wish to raise with the Sub-Contractor.

1.A Any stipulation as to the period/periods when Sub-Contract Works can be carried out on site:[s]

See 1C JS.AD to be between ___1st June 1983___ and ___29th August 1983___ *JS*

Period required by Architect to approve drawings after submission ___4 Weeks___ *JS*

1.B Preliminary programme details[t] (having regard to the information provided in the invitation to tender)

Periods required:

(1) for submission of all further sub-contractors drawings etc. (co-ordination, installation, shop or builders' work, or other as appropriate)[u]

~~Not applicable~~ *AD JS*

~~Not applicable~~ *AD JS*

___2 weeks___ *AD*

(2) for execution of Sub-Contract Works: off-site ___4 weeks___ *AD*

on-site ~~6 weeks~~ *AD*

Notice required to commence work on site ___4 weeks___ *AD*

1.C Agreed programme details (including sub-contract completion date: see also Sub Contract NSC/4, clause 11·1)[v]

CARCASSING
COMMENCE ON 18 MAY 1983
COMPLETE BY 6 JUNE 1983

AD JS

SECOND FIX & 1ST TEST
COMMENCE ON 20 JULY 1983
COMPLETE BY 8 AUGUST 1983

FINAL TEST & LIVEN UP
DURING W/E 10 AUGUST 1983

2. Order of Works to follow the requirements, if any, stated in Schedule 1, item 11[w]

1.20 Conditions for agreement NSC/1, page 9

NOMINATED, NAMED AND DOMESTIC SUB-CONTRACTORS

3.A Attendance proposals (other than †general attendance).[x]

(a)	Special scaffolding or scaffolding additional to the Contractor's standing scaffolding.	Boarded scaffolding for outside lights and ~~dining hall area~~ JS MOBIL TOWER WITH 2m × 1.5m PLATFORM IN DINING HALL AREA AD JS
(b)	The provision of temporary access roads and hardstandings in connection with structural steelwork, precast concrete components, piling, heavy items of plant and the like.	not applicable AD
(c)	Unloading, distributing, hoisting and placing in position giving in the case of significant items the weight and/or size. (To be at the risk of the Sub-Contractor).	none AD
(d)	The provision of covered storage and accommodation including lighting and power thereto.	Weatherproof and securable shed 4m × 3m for fittings with light and power points AD METERED ELECTRICAL SUPPLY WILL BE PROVIDED AND SUBCONTRACTOR WILL BE JS CHARGED FOR POWER CONSUMED AD
(e)	Power supplies giving the maximum load.	for handtools only AD
(f)	Maintenance of specific temperature or humidity levels.	none AD
(g)	Any other attendance not included under (a) to (f) or as †general attendance under Sub-Contract NSC/4, paragraph 27·1·1.	none AD

†Note: For general attendance see clause 27·1·1 of Sub-Contract NSC/4 which states: "General attendance shall be provided by the Contractor free of charge to the Sub-Contractor and shall be deemed to include only use of the Contractor's temporary roads, pavings and paths, standing scaffolding, standing power operated hoisting plant, the provision of temporary lighting and water supplies, clearing away rubbish, provision of space for the Sub-Contractor's own offices and for the storage of his plant and materials and the use of messrooms, sanitary accommodation and welfare materials." See SMM, 6 edn., B.9.2.

1.21 Conditions for agreement NSC/1, page 10

Probably the most important point to ensure when completing page 1 is that the date to be inserted must be the date when tender NSC/1 form is signed as 'approved' by the architect on behalf of the employer. There is reference to this on page 1 of NSC/1 as can be seen in figure 1.19.

The four *recitals* on page 1 of NSC/2 (the 'whereases') are generally self-explanatory; they refer to the procedure as a result of which agreement NSC/2 is to be executed. The fourth recital is interesting in that it 'exempts', so to say, the architect from liability for the contents of the agreement and from liability for the contents of the tender.

NSC/2 conditions A number of points require clarification in the conditions contained in NSC/2.

The actual wording of each clause is shown in small print and these clauses are followed by a brief statement by way of explanation.

Clause 1

1.1 The Sub-Contractor shall, after the Architect has issued his preliminary notice of nomination under clause 35.7.1 of the Main Contract Conditions, forthwith seek to settle with the Main Contractor the Particular Conditions in Schedule 2 of the Tender.

1.2 The Sub-Contractor shall, upon reaching agreement with the Main Contractor on the Particular Conditions in Schedule 2 of the Tender and after that Schedule is signed by or on behalf of the Sub-Contractor and the Main Contractor, immediately through the Main Contractor so inform the Architect . . .

This clause is concerned with the sub-contractor's obligations to follow the procedures (which the main contractor is obliged by SF 80, cl 35 to follow), to reach agreement with the main contractor regarding schedule 2 of NSC/1.

Clause 2

2.1 The Sub-Contractor warrants that he has exercised and will exercise all reasonable skill and care in:

  .1 the design of the Sub-Contract Works in so far as the Sub-Contract Works have been or will be designed by the Sub-Contractor; and

  .2 the selection of materials and goods for the Sub-Contract Works in so far as such materials and goods have been or will be selected by the Sub-Contractor; and

  .3 the satisfaction of any performance specification or requirement in so far as such performance specification or requirement is included or referred to in the description of the Sub-Contract Works included in or annexed to the Tender.

Nothing in clause 2.1 shall be construed so as to affect the obligations of the Sub-Contractor under Sub-Contract NSC/4 in regard to the supply under the Sub-Contract of workmanship, materials and goods.

2.2 .1 If, after the date of this Agreement and before the issue by the Architect of the instruction on Nomination NSC/3 under clause 35.10.2 of the Main Contract Conditions, the Architect instructs in writing that the Sub-Contractor should proceed with
  .1 the designing of, or
  .2 the proper ordering or fabrication of any materials or goods for the Sub-Contract Works the Sub-Contractor shall forthwith comply with the instruction and the Employer shall make payment for such compliance in accordance with clauses 2.2.2 to 2.2.4.

  .2 No payment referred to in clauses 2.2.3 and 2.2.4 shall be made after the issue of Nomination NSC/3 under clause 35.10.2 of the Main Contract Conditions except in respect of any design work properly carried out and/or materials or goods properly ordered or fabricated in compliance with an instruction under clause 2.2.1 but which are not used for the Sub-Contract Works by reason of some written decision against such use given by the Architect before the issue of Nomination NSC/3.

  .3 The Employer shall pay the Sub-Contractor the amount of any expense reasonably and properly incurred by the Sub-Contractor in carrying out work in the designing of the Sub-

Contract Works and upon such payment the Employer may use that work for the purposes of the Sub-Contract Works but not further or otherwise.

.4 The Employer shall pay the Sub-Contractor for any materials or goods properly ordered by the Sub-Contractor for the Sub-Contract Works and upon such payment any materials and goods so paid for shall become the property of the Employer.

.5 If any payment has been made by the Employer under clauses 2.2.3 and 2.2.4 and the Sub-Contractor is subsequently nominated in Nomination NSC/3 issued under clause 35.10.2 of the Main Contract Conditions to execute the Sub-Contract Works the Sub-Contractor shall allow to the Employer and the Main Contractor full credit for such payment in the discharge of the amount due in respect of the Sub-Contract Works.

Clause 2 is generally concerned with design, materials and performance specification.

Clause 2.1 sets out the sub-contractor's design warranty. The warranty comes into effect when the sub-contractor and employer enter into agreement NSC/2 and when the employer (or the architect on his behalf) approves tender NSC/1.

Not until this has been done should design works be carried out or material be ordered prior to nomination and as one of the objects of this arrangement is to provide a framework in which the design *may* be put in hand the importance of the sub-clause will be apparent.

Clause 2.2 makes provision for the sub-contractor to be paid for design work and for materials ordered if no nomination is made. In this event the employer may, upon payment, use the design work and become owner of the materials. If the sub-contractor is *not* prepared to let his design or materials be used by another contractor he should make this clear in his offer to the architect.

Clause 3

3.1 The Sub-Contractor will not be liable under clauses 3.2, 3.3 or 3.4 until the Architect has issued his instruction on Nomination NSC/3 under clause 35.10.2 of the Main Contract Conditions nor in respect of any revised period of time for delay in carrying out or completing the Sub-Contract Works which the Sub-Contractor has been granted under clause 11.2 of Sub-Contract NSC/4.

3.2 The Sub-Contractor shall so supply the Architect with such information (including drawings) in accordance with the agreed programme details or at such time as the Architect may reasonably require that the Architect will not be delayed in issuing necessary instructions or drawings under the Main Contract, for which delay the Main Contractor may have a valid claim to an extension of time for completion of the Main Contract Works by reasons of the Relevant Event in clause 25.4.6 or a valid claim for direct loss and/or expense under clause 26.2.1 of the Main Contract Conditions.

3.3 The Sub-Contractor shall so perform his obligations under the Sub-Contract that the Architect will not by reason of any default by the Sub-Contractor be under a duty to issue an instruction to determine the employment of the Sub-Contractor under clause 35.24 of the Main Contract Conditions provided that any suspension by the Sub-Contractor of further execution of the Sub-Contract Works under clause 21.8 of Sub-Contract NSC/4 shall not be regarded as a 'default by the Sub-Contractor' as referred to in clause 3.3.

3.4 The Sub-Contractor shall so perform the Sub-Contract that the Contractor will not become entitled to an extension of time for completion of the Main Contract Works by reason of the Relevant Event in clause 25.4.7 of the Main Contract Conditions.

This clause deals with delay in supply of information and in performance by the sub-contractor.

It requires the sub-contractor to provide the architect with information (including drawings) in accordance with an agreed programme BUT (and it is a big 'but') this requirement only comes to pass when the architect has issued his instruction on 'Nomination' NSC/3 form.

Clause 4

4 The Architect shall operate the provisions of clause 35.13.1 of the Main Contract Conditions.

NSC/2, clause 4, refers to the procedure by which the architect directs on value of sub-contract work in interim certificates and the information he shall give the sub-contractor which is, incidently, more comprehensive than it has been in the past.

Clause 4 provides the (by this time) *nominated* sub-contractor with a direct right against the employer to ensure that the architect directs the main contractor as to the amounts for the sub-contractor which are included in the amounts stated as due in interim certificates. This reduces the sub-contractor's dependence upon the main contractor in this respect. This is a similar arrangement to that which exists in the 1963 conditions of contract.

Clause 5

5.1 The Architect shall operate the provisions in clauses 35.17 to 35.19 of the Main Contract Conditions.

5.2 After due discharge by the Contractor of a final payment under clause 35.17 of the Main Contract Conditions the Sub-Contractor shall rectify at his own cost (or if he fails so to rectify, shall be liable to the Employer for the costs referred to in clause 35.18 of the Main Contract Conditions) any omission, fault or defect in the Sub-Contract Works which the Sub-Contractor is bound to rectify under Sub-Contract NSC/4 after written notification thereof by the Architect at any time before the issue of the Final Certificate under clause 30.8 of the Main Contract Conditions.

5.3 After the issue of the Final Certificate under the Main Contract Conditions the Sub-Contractor shall in addition to such other responsibilities, if any, as he has under this Agreement, have the like responsibility to the Main Contractor and to the Employer for the Sub-Contract Works as the Main Contractor has to the Employer under the terms of the Main Contract relating to the obligations of the Contractor after the issue of the Final Certificate.

Just as clause 4 was concerned with *interim* payments, clause 5 is concerned with *final* payments. It *provides the employer with a duty to make final payment to the sub-contractor* but it also sets out *obligations on the sub-contractor's part*.

It provides the sub-contractor with a direct right against the employer to ensure that the employer follows the appropriate procedures, set out in clause 35.17 of JCT 80, regarding early final payment in return for which the sub-contractor is obliged *direct* to the employer to rectify faults in the sub-contract works before and after issue of the final certificate under the main contract. Clause 5 should be read in conjunction with clause 35.18.

Clause 6

6 Where the Architect has been under a duty under clause 35.24 of the Main Contract Conditions except as a result of the operation of clause 35.24.6 to issue an instruction to the Main Contractor making a further nomination in respect of the Sub-Contract Works, the Sub-Contractor shall indemnify the Employer against any direct loss and/or expense resulting from the exercise by the Architect of that duty.

Clause 6 places upon the architect a duty to make a further nomination in the event of the sub-contractor making default in respect of a number of matters referred to in clause 29 of NSC/4, in the event of him becoming bankrupt or determining his employment under clause 30 of the same form.

Clause 6 gives the employer a *direct right against the nominated sub-contractor* for an indemnity against the direct loss and/or expense which the employer may incur if the architect, because of some fault, etc, of the sub-contractor, has had to issue a further nomination to carry out and complete the sub-contract works.

Clause 7

7.1 The Architect and the Employer shall operate the provisions in regard to the payment of the Sub-Contractor in clause 35.13 of the Main Contract Conditions.

contract. It is what the medical profession would call preventative medicine.

It follows, then, that as the alternative method leaves numerous matters to be resolved *after* the sub-contractor has been nominated, it is not to be encouraged. The JCT *Guide* reads:

> 'It is suggested that the basic method may be found to be the more convenient arrangement for nominated sub-contract tenders which can be approved by the Employer *before* the Main Contract is let. The preferences of prospective Nominated Sub-Contractors should always be an important consideration in deciding which method to adopt. The alternative method may be considered suitable particularly where the Nominated Sub-Contract Works may not be critical to the progress of the Works as a whole. However, even in these circumstances where the Contractor's programme will have been settled... the scope for a proposed sub-contractor on commencement dates and periods on site, etc, is inevitably limited by the Contractor's programme. It may well be found that ... the basic method would be appropriate ...'

The wording regarding the 'preferences of prospective Nominated Sub-Contractors' is of significance for specialist contractors. Building and installation works which are an integral part of most buildings are almost invariably 'critical to the progress of the Works as a whole'.

There is little doubt but that the basic method of nomination should be used whenever building services, electrical and similar sub-contractors are employed. Indeed, apart from window-box planters and flag-pole installers, it is difficult to think of contractors who could not, with a modest stretch of the imagination, be regarded as 'critical' on some contracts and perhaps if a royal opening ceremony were envisaged even gaily planted window-boxes and operative flag-poles might appear on the critical path!

1.7 Mobilisation

Mobilisation of the project should see a transfer of initiative from employer to contractor and the project manager's tasks should be concerned with that transfer.

Mobilisation by the project manager The checklist for the local authority project managers in the housing department, referred to above, is concerned with reminding consultants of their responsibilities and liaising with other local authority officers such as valuers and estate managers regarding the effect of the contract on their activities.

Other tasks are concerned with advising residents of the project's commencement, ensuring the security of their property and initiating procedures to ensure minimal inconvenience to residents.

There is a need to establish a communications network and an initial meeting is recognised as the most appropriate medium to that end. Agenda for such a meeting might include establishing arrangements to ensure the residents' safety, security and access and that factors affecting adjoining owners and their property are known to all parties concerned with extension of the project.

Agenda should also be concerned with arrangements for monitoring and controlling progress and quality and for interim payments on account.

Responsibilities for arranging agenda, giving notice of meetings, recording decisions and ensuring their implementation should be determined at the initial meeting.

This is the time when contractual, procedural and human relationships are established which effect the whole contract period. It is important to 'get off on the right foot'. Attitudes which might promote separation into 'them and us' groups and confrontation should be avoided.

That said, the contractor should ensure that he will be able to function effectively within the administrative framework proposed by the architect or project manager. He should have positive and clearly stated alternative proposals if those proposed by the architect are unsatisfactory. It is the contractor who is obliged to 'carry out and complete the works' and he is entitled, within the conditions imposed by the contract, to manage the progress of the works in any manner he chooses.

Mobilisation by the contractor is implementation of his pre-tender plan and estimate.

The contractor's estimate and pre-tender programme provide the means of obtaining work and the foundations upon which, if the tender is accepted, the detailed planning and execution of the works are based. If the foundations are not sound the project, like a building, will fail.

Provided the contractor has adopted the methodical approach discussed in section 1.3 for completion of his estimate and its conversion into a tender, he should be adequately prepared to carry out and complete the works.

If the estimate was prepared in accordance with the recommendations of the CIOB Code of Estimating Practice, documentation regarding quotations from sub-contractors and suppliers, the contractor's unit rates, his allowances for preliminaries and other data will be readily available for mobilisation.

In practice, the data will require refinement because the pressures of time during the estimating stage will generally have precluded an exhaustive analysis of methods and available resources. The contractor is aware at that time that his is one of several tenders and that the work of estimating will probably be abortive. His concern at the time of tender is to obtain work rather than prepare for its execution.

Mobilisation is, then, the time when more precise and perhaps more competitive estimates may be obtained from sub-contractors and suppliers and when other methods of construction may be explored with a view to improved performance.

Uniform procedures A measure of uniformity of procedures is an essential aspect of contract administration and the contractor should have a manual of standard procedures for use by all members of the firm.

Such manuals are extensive and outside the scope of this text but typically they will include:

- instructions regarding the operation of procedures
- schedules of proforma to be used for enquiries, requisitions and purchase of materials, sub-contractors' work (labour only and labour and materials), mechanical and non-mechanical plant, suppliers' and sub-contractors' design warranties and performance bonds
- responsibilities for ordering including orders for major, minor and emergency items.

Procedures should be determined for progress and cost monitoring and control to which further reference is made elsewhere in this text.

All procedures should be framed with regard to the conditions of the main and sub-contracts. In addition to the JCT forms of contract which provide the models most generally discussed in this text there are, for example, 'model conditions of contract for the hire, erection and dismantling of scaffolding' approved by BEC and the National Association of Scaffolding Contractors which may be used where appropriate.

Appendix 1A contains the index for an ordering procedures manual, the schedule of purchasing forms and the contents page for materials purchasing for a medium-sized contractor to indicate the scope of such manuals. Without established administrative procedures, which must be followed by personnel at all levels in the organisation, orderly management is impossible. Procedures should, however, be designed so that they are *enablers* rather than *inhibitors*. They should endorse the responsibilities and authority of managers so that all concerned may perform their tasks to achieve the corporate and project aims.

Section 1C
EXECUTION OF THE WORKS

1.8 Carrying out the works

JCT 80 states that the principal obligation of the contractor is to 'carry out and complete the works' and similar wording occurs in the other JCT forms of contracts. The extent to which the contractor is involved in the design of the works varies significantly from form to form but in all the forms the contractor is responsible for 'carrying out', for the 'execution' of the works.

Execution of the works, the building or construction process, is largely concerned with management of the resources necessary to complete the works. The 'resources' are those referred to in the contractor's estimate and tender.

Indeed, in a perfect world, nothing should stand in the way of the progress of the works.

But such a scenario assumes that the employer's requirements were fully explained to the designers who translated them into data which provided the contractor with all he needed to know in order to tender and to carry out and complete the works.

In practice, such a scenario seldom if ever exists for various reasons. Employers seldom know exactly what they need at an early stage in the project's life. Technology may change during the design and construction stages so that the employer's needs change, his proposed tenants may have different requirements from those envisaged, etc, etc. There are many reasons why the works which the contractor tendered to do bears little resemblance to those he has to build.

Most conditions of contract, certainly these published by JCT, recognise the world's imperfections and include conditions to enable the parties to supplement the data provided by the architect as the basis of the contractor's tender, to change the conditions under which the contractor is required to carry out the works or to change the actual works themselves. This section, section 1.7, considers some of the conditions in JCT 80 which facilitate changes and other contingencies.

During this stage the tasks of the project manager are concerned primarily with monitoring the performance of the consultants and contractors to safeguard the employer's interest and ensure his requirements are met.

As the client's representative he is the recipient of the various architect's certificates referred to below. He sanctions changes arising from architect's instructions, reconciles 'actual' with 'planned' expenditure and progress. He sanctions the extensions of time which are the subject of section 2 of this text.

The contractor's responsibility and obligations are contained in clauses 1.4 and 2.1 of JCT 80. These clauses are fundamental to the contract and are quoted in full:

1.4 Notwithstanding any obligation of the Architect to the Employer and whether or not the Employer appoints a clerk of works, the Contractor shall remain wholly responsible for carrying out and completing the Works in all respects in accordance with clause 2.1, whether or not the Architect or the clerk of works, if appointed, at any time goes on to the Works or to any workshop or other place where work is being prepared to inspect the same or otherwise, or the Architect includes the value of any work, materials or goods in a certificate for payment, save as provided in clause 30.9.1.1 with regard to the conclusiveness of the Final Certificate.

2.1 The Contractor shall upon and subject to the Conditions carry out and complete the Works shown upon the Contract Drawings and described by or referred to in the Contract Bills and in the Articles of Agreement, the Conditions and the Appendix (which Drawings, Bills, Articles of Agreement, Conditions and Appendix are in this Contract referred to collectively as 'the Contract Documents') in compliance therewith, using materials and workmanship of the quality and standards therein specified, provided that where and to the extent that approval of the quality of materials or of the standards of workmanship is a matter for the opinion of the Architect, such quality and standards shall be to the reasonable satisfaction of the Architect.

These clauses make the contractor 'wholly responsible for carrying out and completing the Works'. It is the architect's opinion which determines the quality of materials and or the standards of workmanship which are to be to his 'reasonable satisfaction'.

Clause 2.1 states the documents which collectively comprise the 'Contract Documents'.

Whilst the contractor has the responsibility and obligations referred to above the first recital makes it clear that responsibility for the preparation of drawings and contract bills rests with the architect. There is no reference to design being undertaken by consultants other than the architect.

All drawings, schedules, etc, flow through the architect.

The contractor has no responsibility or obligations under clauses 1 or 2 for design.

Furthermore, clause 2.2.1 states that:

'... nothing contained in the Contract Bills shall override or modify the application or interpretation of that which is contained in the Articles of Agreement, the Conditions or the Appendix'.

This clause was introduced at a time when there was a tendency among quantity surveyors to include items in bills of quantities which were, from the contractor's standpoint, more onerous than was the intention of the conditions of contract.

If the contractor finds any discrepancy in or divergence between the contract drawings, the contract bills, or any instruction, drawing or the numbered documents issued by the architect he is obliged to immediately give the architect a written notice and the architect is obliged to issue an instruction. Clause 2.3 makes action by the contractor conditional upon him finding any discrepancy, etc. Should he not find any discrepancy, he is not obliged to take action.

Contract sum The contract sum and adjustments to it are referred to in clauses 3 and 14.

Clause 14 states that the quality and quantity of the work included in the contract sum are to be 'that which is set out in the Contract Bills'. The contract sum may only be adjusted or altered in accordance with the express provisions of the conditions (discussed later in this section) and in the event of an error in the contract bills to which reference is made in clause 2.2.2.2.

Contract sum adjustments Several conditions of contract provide that an amount may be added to or deducted from the contract sum. Such amounts are taken into account in the computation of the next interim certificate following such ascertainment. The contractor is paid such amount as the works proceed. He does not have to wait until completion of the work. It is in the contractor's interest to ascertain any such amounts as quickly as possible because it expedites payment.

He should make available any information the quantity surveyor requires in order to value or ascertain any amounts and persuade him to act. The contractor's positive cash flow is dependent on him receiving prompt payment in compliance with the conditions of contract.

Errors in contract bills and discrepancies in or divergences between documents do not vitiate the contract. The Contract Bills are prepared in accordance with the appropriate Standard Method of Measurement. Should there be an error or omission it is corrected and treated as a variation (cl 2.2). Variations are discussed below.

Architect's instructions (cl 4), must be in writing. Virtually all changes arise as the result of architect's instructions. Clause 4 is, therefore, extremely important because it states the procedures to be used for the issue of instructions.

The contractor is obliged to comply with all such instructions in respect of which the architect is expressly empowered by the conditions of contract to issue.

Provided he makes reasonable objection in writing to the architect, the contractor need not comply with an instruction which requires a variation within the meaning of clause 13.1.2, which clause is concerned with addition to, alteration or omission of any obligations or restrictions imposed by the employer in the contract bills in regard to access to the site, limitations of working space or working hours or the execution or completion of the work in any specific order. Nomination of a sub-contractor to undertake work which is measured and priced in the contract bills is excluded.

The employer is empowered to employ and pay other persons to execute any work which is necessary to give effect to an architect's instruction should the contractor not comply with an instruction within 7 days after receipt of a written notice from the architect.

Upon receipt of what purports to be an instruction issued to him by the architect, the contractor may request the architect to specify in writing the provision of the conditions which empowers the issue of the instruction. The architect is obliged to comply with the request.

If the contractor then complies with the instruction, the issue of it is deemed to have been empowered by the provision of the conditions specified by the architect in answer to the contractor's request.

The procedure if instructions are given 'otherwise than in writing' are stated in clause 4.3.2, which reads:

4.3 .2 If the Architect purports to issue an instruction otherwise than in writing it shall be of no immediate effect, but shall be confirmed in writing by the Contractor to the Architect within 7 days, and if not dissented from in writing by the Architect to the Contractor within 7 days from receipt of the Contractor's confirmation shall take effect as from the expiration of the latter said 7 days. Provided always:
.2.1 that if the Architect within 7 days of giving such an instruction otherwise than in writing shall himself confirm the same in writing, then the Contractor shall not be obliged to confirm as aforesaid, and the said instruction shall take effect as from the date of the Architect's confirmation; and
.2.2 that if neither the Contractor nor the Architect shall confirm such an instruction in the manner and at the time aforesaid but the Contractor shall nevertheless comply with the same, then the Architect may confirm the same in writing at any time prior to the issue of the Final Certificate, and the said instruction shall thereupon be deemed to have taken effect on the date on which it was issued otherwise than in writing by the Architect.

If the contractor and architect delay taking action stated in clauses 4.3.2 and 4.3.2.1 until the latest moment, there is a risk that the works might be delayed.

In order to expedite instructions, the contractor may adopt various procedures. One such is to keep on site a proforma book in which are entered any oral 'instruction' as soon as possible after they have been given. Such 'confirmation' of architect's instruction sheets should be signed by the architect whilst he is still on site.

JCT 80 does not specify the means of transmitting the written instructions. It does not, for instance, state if first or second class post is envisaged.

Nor does JCT 80 recognise the use of Fax or similar transmissions, but Fax is given as a method of service of documents in the JCT Arbitration Rules. Where service is by Fax, for record purposes, the document 'must forthwith be sent by first class post or actually delivered!'

Clause 4.3.2.2, quoted above, provides a procedure in the event of an oral instruction not being confirmed by either or both the architect and contractor at time of issue. The architect may confirm it in writing at any time prior to the issue of the final certificate.

This procedure is occasionally used if an oral instruction is given by the architect which leads to a variation but which is overlooked. The instruction is not in doubt and the contractor acts on it. It is only when the quantity surveyor seeks the authority of an architect's instruction in order to include the variation in the final account that he discovers the oversight.

In most instances confirmation is given by the architect. Not infrequently, however, personnel changes or oral instructions are forgotten and the quantity surveyor has no authority to include the variation.

The contractor is wise not to rely on retrospective confirmation of architect's instructions but to ensure that they are confirmed before he takes action because he is at risk of not being reimbursed if he acts on an oral instruction.

Documents and certificates are fundamental to building contracts. The contractor is imprudent to enter into a contract without documents. Reference has been made, above, to the requirements for architect's instructions to be in writing. Sam Goldwyn's remark (page 27) is pertinent to building works.

Administration of documents Major construction projects involve the production and use of thousands of documents (including drawings) by numerous people. The management of information on that scale requires complex systems. Expert advice should be sought regarding an appropriate system. Computer programs are readily available but some projects may make special demands.

The administration of documents for smaller projects may be designed for operation manually.

Large or small, the installation and operation of an appropriate system necessitates the identification of the following information in order to specify criteria and parameters and thus design a system for administration of the documentation:

- purpose of the information contained in the documentation
- format and nature of the documentation (paper, size, film, disc, tape, etc)
- methods of transmission (manual, postal service, courier, Fax, etc)
- initiators (designer, specifier, supplier, installer, etc)
- users of the information contained in the documents
- location of initiators ⎫
- location of users ⎬ relative locations may influence transmission method.
- time constraints (critical, non-critical)
- dependency and/or interdependency of information
- security constraints (to determine transmission method/s)
- period for which documents are to be retained
- extent to which they may need to be consulted.

Typically, the preparation of an information schedule requires identification of a number of features. For example:

- a code/reference
- the name of the feature
- a term which identifies it
- its characteristics
- the source of input or form of output.

For planning specifications the entries under the above headings might read:

- a short code, perhaps a mnemonic or acronym)
- specification
- for each operation in the project
- dimensions and quantities of materials or description of work to be done
- from detailed planning routine.

The tendency for the construction stage of some construction projects to commence before the design is complete makes it necessary for the contractor to retain, documents which record changes that may have occurred in design or employer's requirements.

Earlier issues of drawings which, for example, have been revised, may be of no interest to the site manager, whose concern is with managing construction, but they will be important to the surveyor who is concerned with the cost implications of such changes.

Obsolete documents must, therefore, be retrievable at least until the contractor has been paid in full. If he has any responsibility for design, documents must be retrievable until the time limit of his responsibility has expired.

Co-ordinated project information (CPI) A move towards the co-ordination of project information commenced with the publication by the Building Project Information Committee (BPIC), in 1987, of guides to the preparation of specifications, drawings and common arrangement of work sections. The common arrangement is compatible with SMM7 and with the National Building Specification.

CPI documentation suggests that there is no single 'best' arrangement for drawings which will apply to all projects. There are many factors such as size and complexity of project and types of construction with the project which will influence the choice. The most effective arrangement will result from giving the right emphasis to each factor for the particular circumstances.

Reaearch for BPIC identifies three 'arrangements' for drawings.

The terms used in figure 1.23 are defined as follows:

Non-systematic arrangement
No groups of related drawing sheets are apparent from either the drawing numbering or the drawing types. A typical example is a set of about 30 drawings which, although their preparation may have been given some logical thought, appear as a mixture of layouts, details and schedules.

Simple systematic arrangement
Groups of related drawing sheets are apparent from the drawing numbering and/or the drawing types. A typical example is a set of about 100 drawings which is separated into location drawings, schedules, assembly drawings and component drawings.

Advanced systematic arrangement
Initial grouping is similar to the simple systematic approach but further, more detailed, division is apparent from the drawing numbering. A typical example is a set of more that 250 drawings separated into location, coordination, schedule, assembly and component drawings further divided by parts of the building, eg foundations, roof, floors.

Small projects
Typically of small to medium size and of simple construction. Requiring a minimum of information management.

Medium projects
Typically of small to medium size but complicated construction or of large size but simple construction. Requiring a fairly simple but systematic approach to information management.

Large projects
Typically of large size and complicated construction. Requiring an advanced form of information management.

A project which corresponds to changeover point A in figure 1.23 might, for instance, be a hostel with a floor area of about 1,000 m^2, while one at changeover point B could be university teaching accommodation, a hospital or a building of similar complexity in excess of about 10,000 m^2 floor area.

Choice of arrangement The essence of good arrangement is the division of the whole set of production drawings into easily recognisable groups. The choice of grouping is affected by the following factors.

Type of project The typical small project will need no more elaborate grouping than the separation of general arrangements from details.

The typical medium project will need no more elaborate grouping than by type of information, eg division into location, assembly and component drawings.

The typical large project will need group-

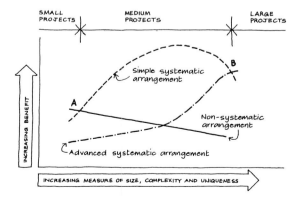

1.23 *The effect of drawings arrangement*
BPIC, 1987

ing by type of information and further division by, for instance, parts of the building. Special groups such as coordination drawings may also be required.

Some projects will display conflicting characteristics and will be difficult to define as small, medium or large. In these cases drawings groups should be chosen to accommodate the characteristics. This means that some small projects will need grouping by type of information and some medium projects will benefit from subdivision of the assembly group. Some medium and most large projects will need grouping by zones or blocks. Projects having a variety of small units (eg housing sites or factory unit developments) may need grouping by unit types.

In the JCT 80 contractual relationship the contractor seldom has control over the choice of arrangement but the principles described above may be of assistance to him if, for example, he is involved in a design-and-build contract or when arranging drawings for his own production planning purposes.

Documents and certificates are the subject of clause 5 which is concerned with their custody, issue and return.

Custody of contract documents The (original) contract drawings and the contract bills remain in the custody of the architect or quantity surveyor so as to be available at all reasonable times for inspection by the employer and contractor (cl 5.1). The architect is obliged to provide the contractor, without charge, immediately after examination of the contract with:

- one copy certified on behalf of the employer of the contract documents
- two further copies of the contract drawings and of the unpriced bills of quantities (cl 5.2).

Descriptive schedules and master programme As soon as possible after execution of the contract there is, without charge, an exchange of documents in that the architect is to provide the contractor with 2 copies of any descriptive schedules or other like documents for use in carrying out the work and the contractor is to provide the architect with 2 copies of his master programme. The contractor is obliged, within 14 days of any decision by the architect under clause 25.3.1 or 33.1.3 to provide him with 2 copies of any amendments and revisions to take account of the architect's decision (cl 5.3.1.2).

Clause 5.3.1 states that nothing contained in the descriptive schedules or other documents (nor in the master programme) referred to above shall impose any obligation beyond those imposed by the contract documents.

So what if the descriptive schedules etc appear, in practice, to impose further obligations?

Such documents might be regarded as being covered by clause 5.4 which provides that:

'as and when from time to time may be necessary the Architect without charge to the Contractor shall provide him with 2 copies of such further drawings or details as are reasonably necessary either to explain and amplify the Contract Drawings or to enable the Contractor to carry out and complete the Works in accordance with the Conditions'.

In practice it is unlikely that 2 copies of drawings and documents will be sufficient for the contractor's needs. Further copies should be obtained from the architect, for which he would be entitled to charge, or the contractor may make copies for his own use and for the use of sub-contractors.

If the contractor discovers any discrepancy in or divergence between the descriptive schedules etc provided to 'explain and amplify...' and the contract drawings or the contract bills he should immediately give the architect a written notice specifying the discrepancy, etc, as required by clause 2.3. The architect is then obliged to issue an instruction regarding the action to be taken by the contractor. For example, which of the drawings to use.

Clause 5.5 requires the contractor to keep one copy of the contract drawings, one copy of the unpriced bills of quantities and one copy of the descriptive schedules, etc, and one copy of 'further drawings ... [in effect, a copy of all the current documents, programme and drawings].... so as to be available to the architect or his representative at all reasonable times'.

Clause 5.5 does not state where they are to be available. An office close to the construction site may be the most suitable location.

Upon final payment the contractor is obliged, if so requested by the architect, to return to him all drawings, etc, which bear his name (clause 5.6).

Clause 5.7 limits the use of the documents to the contract. Neither the employer, the architect nor the quantity surveyor may 'divulge or use except for the purposes of this Contract any of the rates or prices in the Contract Bills'.

The concluding clause, 5.8, obliges the architect to issue any certificates required by the conditions of contract to the employer and, immediately, to send a duplicate copy to the contractor.

Variations are the subject of clause 13. The subject has been brought forward in this text because several clauses make reference to them.

The term 'variation' is defined as meaning an alteration or modification of the design, quality or quantity of the works as shown upon the Contract Drawings and described by or referred to in the Contract Bills including:

- the addition, omission or substitution of any work
- the alteration of the kind or standard of any of the materials or goods to be used in the works
- the removal from the site of any work executed or materials or goods brought on to the site by the contractor which are not in accordance with the contract
- the addition, alteration or omission of any obligations or restrictions imposed by the employer in the contract bills in regard to:
    - access to the site or use of any specific parts of the site
    - limitations of working space
    - limitations of working hours
    - the execution or completion of the work in any specific order.

Variations do not include nomination of a sub-contractor to supply and fix materials or goods or to execute work in the bills of quantities (cl 13.1).

Causes of variations Variations arise from architect's instructions or may be sanctioned in writing by the architect. The contract is not vitiated by the issue of variations (cl 13.2).

A Variation is, then, a *change* from what the contractor contracted to do. Indeed, *change* is the term used in USA and in CD 81.

Provisional sums are sums included in bills of quantities to be expended by the contractor on instructions from the architect. Architect's instructions in this respect may require a variation (cl 13.3).

Valuation of variations Clauses 13.4 and 13.5 are concerned with valuing variations. The quantity surveyor is obliged to value variations of the contractor's work 'in accordance with the provisions of clause 13.5'.

Variations of nominated sub-contractor's work is valued in accordance with the relevant provisions of Sub-contract NSC/4.

If the contractor tenders for work which involves the expenditure of a provisional sum, the work is valued in accordance with the accepted tender.

Valuation rules The rules identify characteristics which determine the way in which the varied work is valued:

- *work which is of similar character* to, is executed under similar conditions to, and does not significantly change the quantity of work set in the contract bills *is valued using the rates and prices in those bills*
- *work which is of similar character* to work set out in the contract bills but *is not executed under similar conditions* and/ or significantly changes the quantity of

work is *valued using the rates and prices for the work as the basis* for determining the valuation. The valuation includes a fair allowance for the differences in conditions and/or quantity
- *work which is not of similar character* to work in the contract bills *is valued at fair rates and prices.*

The quantity surveyor exercises increasing professional judgement as the varied work moves from being 'of similar character' to '*not* being of similar character'. He should adhere as closely as possible to the rates and prices in the contract bills for similar work, use adjusted (what are often referred to as pro rata or 'star') rates for work not executed under similar conditions and daywork and a similar basis for work which is not of similar character.

The contractor should take particular care to ensure that the 'vouchers' (often referred to as daywork sheets) which record the workmen's names, the plant and materials used are delivered to the architect or his authorised representative for verification 'not later than the end of the week following that in which the work has been executed'.

Clause 13.5.5 refers to a provision that is not always appreciated but which gives the quantity surveyor considerable discretion. The clause states that if the variation 'substantially changes the conditions under which any *other* work is executed, then such other work shall be treated as if it had been the subject of an instruction of the Architect'.

It may be necessary for the contractor to demonstrate that substantial changes to the conditions have occurred because the quantity surveyor does not have such ready access to cost and production data as the contractor.

In this connection the contractor's right to be present at the time of measurement by the quantity surveyor (cl 13.6) may be a useful means of arriving at the facts and figures.

Administering variations Variations not infrequently cause disruption of the works, delay to progress and loss to the contractor because he is unable to demonstrate that variation has occurred. It is, therefore, important that the contractor has administrative procedures which safeguard his interests.

The procedures must enable the contractor to identify, record, measure and value variation.

Identifying variations In order to identify variation it is necessary to know what the contractor contracted to do. Under a JCT 80 contract the works are 'shown upon the Contract Drawings and described ... in the Contract Bills' (cl 2.1).

The contractor prepared his estimate on the basis of that information. If the works he is later required to carry out vary from the contract drawings and bills he is entitled to have the contract sum adjusted (cl 3).

His procedures must, therefore, include mechanisms for comparing the Contract Drawings with those provided subsequently by the architect. It is the 'further drawings and details [provided] to explain and amplify the Contract Drawings or to enable the contractor to carry out and complete the Works' (cl 5.4) which may give rise to variations and which the contractor must identify.

The revision letters on the drawings should assist with the identification but not all changes are recorded in the legend on revised drawings.

Recording variations All variations identified should be recorded indicating the nature of the variation and date when the instruction, further drawing or detail, etc, are received.

The size and nature of the project may influence the method of display. If CCPI system is used the method will be prescribed to some extent, at least.

The contractor will probably prepare his record by operation or activity using a bar chart on which is recorded the date by which information is required. An indicative format is shown in figure 2.1.

The problem of administering variations

```
┌─────────────────────────────────────┐
│ DESIGN VARIATION                    │
│ Distribution: Client   Struct Eng   │
│               Architect Services Eng│
│               QS                    │
│ Proposed change                     │
│ .....                               │
│                                     │
│ Initiated by ......    date ...     │
│ Effect on programme                 │
│ .....                               │
│                                     │
│ Architect .....        date ...     │
│ Effect on cost                      │
│ .....                               │
│                                     │
│ Variation can/cannot be inc in BQ   │
│ QS .....               date ...     │
│                                     │
│ Abortive work involved by:          │
│ .....                               │
│ Fee implication .....               │
│                                     │
│ Decision                            │
│ .....                               │
│                         date ...    │
└─────────────────────────────────────┘
```

1.24 Design variation sheet
  BPIC, 1987

is recognised by BPIC in relation to communication within the design team. Figure 1.24 illustrates a 'design variation sheet' which is intended to assist the design team by making sure that the whole design team is aware of all changes and that they assess the cost and time implications.

The proforma is intended for use during preparation of production information. It is initiated by whoever suggests a change and supported by a drawing where appropriate.

The contractor may well use a similar proforma to record the implications of variations.

Statutory obligations Reference is made above to the need for a contract to be for a lawful purpose. The parties to a contract must have regard to the law but they may agree between themselves on who shall do what to ensure compliance with legal obligations. Clause 6 refers to the agreements between employer and contractor regarding compliance with statutory obligations, notices, fees and charges. It is the contractor who takes the initiative.

He is obliged to comply with and to give all notices required by any Act of Parliament or regulation or bylaw of any local authority or of any statutory undertaker which has jurisdiction with regard to the works.

If he finds any divergence between the statutory requirements and any of the documents or instructions, issued by the architect he is obliged to give to the architect a written notice specifying the divergence.

The architect has 7 days to issue an instruction in relation to the divergence. If the architect's instruction requires a variation it is dealt with under clause 13, Variations, to which reference is made above (cl 6.1).

In the event of action being necessary to deal with an emergency the contractor is required to undertake limited work as is 'reasonably necessary to secure immediate compliance with the Statutory Requirements' (cl 6.1.4.1).

He should then inform the architect of the emergency and of the steps that he is taking. The emergency works are treated as if they arise from an architect's instruction requiring a variation (cl 6.1.4.3).

Provided the contractor carries out the work referred to above in compliance with the contract documents he is not liable to the employer if the works do not comply with the statutory requirements (cl 6.1.5).

The second section of clause 6 is concerned with the payment of fees and charges which are 'legally demandable'. In clause 6.2 the contractor undertakes to pay such fees and to indemnify the employer against liability.

Finally, as far as statutory obligations are concerned, clause 6.3 states that statutory undertakers – public service bodies who have a statutory obligation to provide power, water, etc – 'shall not be sub-contractors within the terms of this Contract' when 'executing such work solely in pursuance of its statutory obligations'. Clause 6.3 refers to clauses 19 (assignment and sub-contracts) and 35 (nominated sub-contractors and suppliers) to which reference is made below.

Such undertakers may be employed as sub-contractors if the work is *not* 'solely in pursuance...'. Thus a gas board could be employed as statutory undertaker bringing in the mains but a sub-contractor for the gas installation and carcasing.

Positioning the buildings in 'three dimensions' is the responsibility of the contractor and he 'shall at no cost to the Employer amend any errors arising from his own inaccurate setting out'. It is, however, the responsibility of the architect to:

'determine any levels which may be required for the execution of the works, and he shall provide the Contractor by way of accurately dimensioned drawings with such information as shall enable the Contractor to set out the Works at ground level'.

It is not unknown for the architect to provide 'pegs' indicating the corners of the proposed building.

Such 'physical' information should not be used as an alternative to the provision of 'accurately dimensioned drawings'. Pegs and profiles may be disturbed by site works and, in any event, there is no permanent record of where the building/s should be positioned should a dispute arise at a later day.

An alternative to the contractor amending any errors is for the employer to condone the errors. In this event the architect 'may instruct that such errors shall not be amended and an appropriate deduction for such errors not required to be amended shall be made from the Contract Sum' (cl 7). The basis for valuing such 'non-amendments' is not stated and may be difficult to determine.

Materials, goods and workmanship The procedures to be adopted to ensure that materials, goods and workmanship conform to description, testing and inspection are contained in clause 8. With regard to materials and goods *standards*, clause 8.1 states that all are to be 'so far as procurable' of the kinds and standards described in the Contract Bills. They are to be 'to the reasonable satisfaction of the Architect'.

Workmanship is more difficult to specify but it is to be 'of the standards described in the Contract Bills' or, if standards are not stated in the bills, 'shall be of a standard appropriate to the Works'.

Whichever of the above 'standards' is used the workmanship is to be 'to the reasonable satisfaction of the Architect'.

As far as materials, goods and workmanship are concerned clause, 8 refers to clause 2.1 which is quoted in full, above.

Insofar as it is possible for the architect to *prove* that standards are complied with, clause 8.2.1 obliges the contractor to provide him, if requested, with vouchers 'to prove that the materials and goods comply with clause 8.1'.

Reasonable time with regard to *executed work* The architect is obliged to 'express any dissatisfaction within a reasonable time from the execution of the unsatisfactory work' (cl 8.2.2).

Inspection and tests Not all questions of standards rest on the opinion of the architect. Some standards can be measured or tested. In many respects standards which can be quantified and/or tested numerically are more satisfactory than those which rest on opinion.

The architect has the power to issue instructions requiring the contractor to open up for inspection any work covered up. He can then arrange for or carry out any test of materials, goods or executed work (cl 8.3). If the inspection reveals that the works are in accordance with the contract the cost of opening up or testing and making good are added to the contract sum.

This brings us to what happens if the work or materials are not in accordance with the contract. This, to a considerable extent, depends on the architect who may:

— issue instructions in regard to removal of the non-complying items and/or
— after consultation with the contractor (who is obliged to consult with any relevant nominated sub-contractor) and with the agreement of the employer,

write to the contractor allowing the items to remain and/or
- after consultation with the contractor (who again, is obliged to consult with any relevant nominated sub-contractor) issue an instruction requiring a variation *But no addition to the contract sum is allowed and no extension of time given* and/or
- having regard to the Code of Practice appended to the conditions, issue instructions under clause 8.3 to open up for inspection or to test as are reasonable to establish his reasonable satisfaction the likelihood or extent of further similar non-compliancies. Whatever the results of the inspection or tests no addition to the contract sum is made. The contractor is entitled to an extension of time unless the inspection or tests reveal that the works were not in accordance with the contract (cl 8.4).

The detail of clause 8.4 has been simplified, above, for purposes of explanation but it means, in effect, that the parties are provided with a number of alternative courses of action depending on the extent of the non-compliance.

Exclusion of persons by architect The architect has the power to issue instructions requiring the exclusion from the works of any person employed thereon, provided he does not act unreasonably or vexatiously (cl 8.5).

It is difficult to envisage circumstances which might lead to the architect exercising powers given in this clause but the provision is there should he wish to do so.

Royalties and patent rights The contractor is obliged to pay all royalties or other sums payable in respect of the works described in the contract bills and to indemnify the employer from all claims arising from the contractor's infringement.

If the matter of royalties arises from an architect's instruction the contractor is not liable in respect of infringement and the costs in connection with the royalties, etc, are added to the contract sum (cl 9).

Access for architect to the works The architect and his representatives have at all reasonable times access to the 'works' (the site itself) and to the workshops or other places where the contractor and/or sub-contractors prepare work for the contract. The contractor is obliged to insert a term in sub-contract conditions which ensures the architect access to their works (cl 11).

The contractor's 'person-in-charge' is the term used in clause 10 to identify the person who might be more usually known by a title such as site manager, agent, contract manager, general foreman or project manager. In section 5 the title 'project manager' refers to the manager who represents the client and has overall responsibility for the entire procurement rather than to the contractor's person-in-charge.

The contractor is obliged to constantly keep upon the works a competent person-in-charge and any instructions given to him by the architect or directions given to him by the clerk of works are deemed to have been issued to the contractor (cl 10).

The contractor should have this clause in mind when designing his administrative procedures. The person-in-charge is, in effect, the contractor's agent and his relationship with the contractor has much in common with the relationship between employer and architect.

It will be noted that the person-in-charge is 'constantly ... upon the Works'.

The clerk of works (one of the most ancient titles in the construction industry) has the duty to act solely as inspector on behalf of the employer under the directions of the architect. The clerk of works, if any, is appointed by the employer and the contractor is obliged to afford the clerk of works every reasonable facility for the performance of his duty.

The clerk of works may give 'directions', (not instructions), which are of no effect unless given in regard to a matter in respect of which the architect is expressly empowered by the conditions to issue instructions and unless confirmed in writing by the architect within 2 working days of the direction being given. If the direction is confirmed then as from the date of issue of the confirmation it is deemed to be an architect's instructions (cl 12).

The contractual position of the clerk of works is not always understood by the incumbent of the post. It is in some respects an invidious role.

Whilst the contractor is obliged to 'afford (the clerk of works) every reasonable facility', the clerk of works directions are of no effect *until* they are confirmed and become architect's instructions.

In practice this condition requires considerable common sense and tact if it is to be effective. Without goodwill it may cause unnecessary delay.

If, for example, the clerk of works considers during an inspection that the trench bottoms are unsuitable he might direct the contractor not to place concrete. In this event the works might be delayed for two days pending the architect's, confirming, instruction.

In practice common sense usually prevails and local arrangements are made to expedite progress but the contractor should have in mind that he may be vulnerable to loss and liquidated damages should amicable local arrangements not be honoured at a later date.

The contractor may, too, find it necessary to commission building services, connect up drains, etc, in advance of the regular progress of his work. He should seek instructions from the architect before undertaking works which would involve him in additional cost. Such works should then be treated as variations.

The treatment of defects in the possessed part during the defect liability period is the same as that described above for the building as a whole.

Injury to persons and property The contractor accepts liability under clause 20, for injury to persons or property and he indemnifies the employer against any expense, liability, loss, claim or proceedings arising under any statute or in common law in respect of personal injury or the death of any person arising out of or in the course of or caused by the carrying out of the works, except to the extent that the same is due to any act or neglect of the employer or person for whom the employer is responsible.

The contractor accepts similar liability to that described above in respect of injury or damage to property 'real or personal' and he indemnifies the employer against any expense, liability, etc.

The reference to 'property real or personal', referred to above, does not include the works, work executed and/or site materials up to and including the date of issue of the certificate of practical completion or up to and including the date of determination of the employment of the contractor (cl 20.3.1).

If the employer has taken partial possession, the possessed part is not regarded as 'the works' or 'work executed' for purpose of clause 20.3.1.

Clause 20, above, states the liabilities the contractor accepts and the indemnity with which the employer is provided. How can this indemnity be assured?

Insurance against injury to persons and property Clauses 21 and 22 are concerned with the steps taken under JCT 80 to provide insurance. Section 3 of this text discusses the obligation taken by the parties with regard to insurance.

Interim certificates and payments Obtaining payment for work in progress is an important management task. Clause 30 contains the procedures and rules for the valuation and certification of work in progress.

The clause is a 'manual' for the contractor in that it sets out procedures for the preparation of interim certificates. Interim payments to the contractor depend on the architect issuing certificates stating the amount due to the contractor from the employer. The certificates are issued 'at the Period of Interim Certificates specified in the Appendix'. The period is one month if none other is stated.

The 'amount due', referred to above, is ascertained by interim valuations made by the quantity surveyor. The quantity surveyor is responsible for the valuation but he normally consults the contractor regarding the ascertainment of some of the amounts which make up the total of the 'amount due'. It is in the contractor's interest to assist the quantity surveyor by providing information regarding costs, etc.

The architect issues a certificate to the employer stating the amount due to the contractor. The contractor is entitled to payment within 14 days from the date of issue of the certificate.

Clause 30.2 describes in detail how the amounts are to be ascertained. The RICS publish proforma, intended for use by quantity surveyors, which follow the format given in clause 30.2 and facilitate preparation of interim valuations.

The architect may include the value of materials which are intended for incorporation in the works *but are not yet delivered to the site* provided they are manufactured and subject to several conditions stated in clause 30.3.

Clauses 30.4 and 30.5 contain rules for ascertainment and treatment of retentions.

The contractor's financial management procedures for projects should include interim accounting systems which use the 'rules for ascertainment of amounts due' and 'treatment of retention' etc (described in the conditions of contract) as models. The contractor's interim accounting system should be used as a 'shadow' of the quantity surveyor's interim valuations.

Clause 30 provides a framework for financial monitoring and control of the project.

Preparation of interim valuations of the work executed at regular (usually monthly) intervals for the purpose of obtaining payments on account is an activity which may be used by the contractor for other purposes.

One such use is the preparation of cost: value reconciliations which are discussed in section 1.11. These reconciliations are an essential aspect of interim cost monitoring and control for the contractor.

Taxation – Value Added Tax and Finance (No. 2) Act 1975 Taxation provisions are stated in clauses 15 and 31.

Clause 15 defines 'VAT Agreement' as the Finance Act 1972 and states that 'any reference in the Conditions to Contract Sum' should 'be regarded as such sum exclusive of any tax and recovery by the Contractor from the Employer of tax properly chargeable . . . .'.

Provision for possible exemption from VAT is contained in clause 15.3 which states that should the supply of goods and services to the employer, after the date of tender, become exempt from VAT, the 'Employer is obliged to pay the Contractor an amount equal to the loss of credit on the supply to the Contractor of goods and services which contribute to the works' (cl 15.3).

Clause 31 refers to the Finance (No. 2) Act 1975 and the Income Tax (Sub-Contractor in the construction industry) Regulations 1975 SI 1960. An aspect of clause 31 which has to be determined is the identity of the 'contractor' because either employer or contractor may be 'contractor' for the purpose of the Act. The employer and contractor should ensure that a deletion is made in the appendix to clarify their intention. The wording in the appendix is:

'Employer at Date of Tender is a "contractor"/is not a "contractor" for purposes of the Act and the Regulations.'

The deletion which should be made, above, is obvious.

Having established the identity of the contractor, clause 31 is largely concerned with operation of the 1975 Regulations which involves obtaining tax certificates if the contractor is not certificated.

There is specific reference in clause 31.9 to arbitration in the event of any dispute or difference arising.

Works by employer or persons employed or engaged by employer Works of the kind contemplated in this clause may arise because the work is described in the contract bills and the contractor permits its execution or because, although the work is not described in the contract bills, the employer requires the work and the contractor consents to its execution by the employer or persons employed by him. The contractor is obliged not to unreasonably withhold his consent (cl 29.1 and 29.2).

Such persons are not sub-contractors. The employer is responsible for them (cl 29.3).

The contractor's relationship with (what we will refer to as) these 'persons' is difficult to define. He is not responsible for them but he permits them to execute the work described in the contract bills or he consents to the employer arranging with the persons for the execution of the work.

The attendance on and/or works to be carried out by the contractor in connection with the persons' work is usually described in the contract bills so that the contractor is able to price it.

Not infrequently the nature and extent of the works is not known at the time when the bills of quantity were prepared so there is no description of attendance or of associated works. In this event the contractor should keep comprehensive records of all the costs he incurs in connection with the persons so that the work, attendance, etc, may be valued and the amount added to the contract sum.

He should, too, obtain an architect's instruction to carry out the work or, alternatively, enter into an agreement with the persons that they will pay the contractor. Such an agreement would be separate from the contract between employer and contractor.

Group decisions and communications
Reference is made in section 1.11, when discussing corporate organisation structure, to the roles of the group board *directing* activities and of the *management* of the divisions which are the source of revenue for the group.

Formal lines of communication for management are provided through the reporting system discussed in section 1.11.

Meetings In the context of construction contracts meetings provide a medium for

---

(i) Giving information and/or instructions to others
(ii) Exchanging information between all those present
(iii) Generating and exchanging ideas
(iv) Formulating for decision-taking
(v) Marketing and sales promotions
(vi) Social purposes
(vii) Encouraging staff participation and involvement in company development

---

1.25 Aims of meetings

communicating information between client, designers, constructors and installers, for generating and exchanging ideas, for formulating for decision-taking and for giving instructions. Figure 1.25 lists the aims of meetings.

Meetings are a valuable medium but they may be a costly and time-consuming device.

Convening a meeting The person convening a meeting should have clearly defined aims. Figure 1.26 provides a checklist for the convener.

It should be a simple matter for the manager to answer the questions in the checklist if he has the Aims of Meetings in front of him.

(a) Have I a clear purpose in mind?
(b) If so, is a meeting the best way of tackling the situation?
(c) Even if it is the ideal way, is there another, more economical, less time-consuming way of tackling the situation?
(d) If a meeting is essential, who should attend?
(e) Should there be a record of the meeting? How?
(f) Should other persons, not attending, be informed of what happened or what was decided?
(g) Have I all the data I need in order to contribute to best effect?
(h) Have the other persons who will be attending sufficient information in advance of the meeting to enable them to contribute to best effect?
(i) Is the proposed venue equipped for audio/visual presentations which may be necessary?

1.26 Checklist for meeting convenor

Experience suggests that major actions are rarely decided by more than four people and that the usefulness of any meeting is in inverse proportion to the attendance. A large meeting, appropriate for aims (ii) or (iii) may, however, be useful for generating and developing ideas on which to base decisions. The answer to question (e) will almost certainly be 'yes' if the aim of the meeting is decision-making. A record of the meeting is likely to be necessary if it was decided to restrict attendance at the meeting and the answer to question (f) would, therefore, probably be 'yes'.

The convener of the meeting will need to be well informed if the meeting aims are in categories (i), (ii) and (iv) (see Q(g)), and others attending should be well informed, probably be means of discussion papers circulated in advance of the meeting.

Types of meeting Meetings vary in type from highly formal parliamentary, central and local government councils, boards, committees, sub-committees and working parties to trade and professional conferences, company 'board meetings' to comply with statutory requirements and meetings of directors or partners to decide policy. In addition, most partners, directors and managers of organisations with which the manager is likely to be concerned usually hold regular meetings for the purpose of organising, monitoring and controlling the progress of their projects and their profitability.

Site meetings are a well-established phenomenon in the construction industry which, some say, are wasteful of resources, ineffective and should be abolished. Sub-contractors and quantity surveyors probably waste more time than most other parties because they are expected to attend site meetings.[6]

Given adequate and timely information the contracting team should be able to carry out and complete the works without the need for site meetings with the designers. It follows that site meetings should be of less importance for traditional contracts (for which the design should be developed before work commences) than for non-traditional contracts where the design is developed in parallel with construction. For traditional contracts the adoption of site meetings is a symptom of poor information flow from design team to those who have to construct. The use of site meetings, as a means of reporting progress, is also questionable. There are ways of reporting progress which do not involve between 5 and 25 people spending a minimum of half a day attending on site. It has been said that: 'the more time you spend in reporting on what you are doing, the less time you have to do anything. Stability is achieved when you spend all your time doing nothing but reporting on the nothing you are doing.'

Whilst report-making is unlikely to monopolise a manager's time it is possible for him to spend a disproportionate amount of his time on non-productive paperwork of report-making type. It has been suggested that if more than one-third of one's time is spent at meetings something is wrong.

Group management methods applied to construction contracts To illustrate the application of group management methods to construction contracts imagine a medium-sized construction division with day-to-day operations managed by a small management group chaired by the contracts director. The group's tasks are concerned with obtaining work, organising, monitoring and controlling progress of contracts and ensuring positive cash flow and profitability. Regular programmed meetings provide the medium for group management, for the passage of information, generating and exchanging ideas, etc, as indicated in figure 1.25.

There are three phases in the progression of meetings:

(a) pre-meeting arrangements
(b) conducting the meeting
(c) implementing decisions taken.

Pre-meeting arrangements; agenda Having established, above, that the aims of the proposed meeting/s must be clear and well defined, the matters to be discussed should be listed as agenda. As a series of regular meetings is planned a standard agenda should be prepared.

The items of the agenda will be related to the group's tasks.

Typical agenda might read:

1. apologies for absence
2. minutes of last meeting
3. matters arising (from the minutes of the previous meeting)
4. enquiries from prospective clients
5. tenders for submission during next week
6. any other business.

Meeting frequency For practical purposes meetings are normally cyclical (weekly, monthly or annual). A regular day of the week and time of the day should be arranged for the meetings which members will 'keep clear' in their diaries. In the absence of pre-arranged meetings members must be given adequate notice.

Each agendum needs attention at frequent intervals because dealing with enquiries from prospective clients and submitting tenders must be matters of some urgency. The meetings should, therefore, be held at weekly intervals. In addition to the items listed above the group needs to review the department's finances and the progress of projects, but these matters may not justify weekly consideration other than in exceptional circumstances. Once a month, however, the group's agenda might be extended by the addition of the following items:

6. financial review
7. project review
8. accounts for settlement
9. any other business.

Attendance at meetings For reasons of economy the number of persons attending the meetings should be kept to a minimum consistent with efficiency. If the management group comprises the contracts director, the senior estimator, the company/departmental secretary, the senior contracts manager/engineer and the senior surveyor, it may be appropriate for the first three persons listed above to attend the weekly meetings but for the group to be extended to include the senior contracts manager/engineer and the senior surveyor for the monthly meetings.

Conducting the meetings Agenda provide the framework for the meeting.

The chair The contracts director will normally 'take the chair' and act as the chairman for the meetings in the example. His duties being:

(a) to guide the members through the agenda and ensure that they do not digress or exceed the time available for the meeting
(b) to ensure that each item is debated adequately and that all members have a fair hearing
(c) to decide the order in which members should speak
(d) to stimulate discussion
(e) to bring the members to a conclusion on each item which should be expressed as an unambiguous resolution. The resolution should identify the person/s who are to take action
(f) ensure that the resolution is accurately and concisely recorded, unless it is decided that a record is not required.

The chair of a management group such as that referred to above seeks the views of members of the group which would help to shape decisions but ultimately the responsibility for decision-making rests with the chairman who, as contracts director, is accountable to the board of directors for his department's performance.

Decisions of democratically constituted councils, boards and committees, such as those of local authorities or professional institutions, are made on a majority vote basis and the committee has corporate responsibility. But for both democratic committees and groups where the chair has ultimate responsibility, the chair should be seen to be fair to all members. It is a difficult role, which leaves the chair open to charges of being 'dictatorial' or 'indecisive' (frequently both at the same time, but by different members of the group). Meanwhile, he has probably regarded himself as the servant rather than the master of the group.

Membership Members should assist the chair to accomplish his duties. All discussion should be addressed to the chair or the meeting will deteriorate into a chat shop. Only one member should speak at a time and members should be restricted as to the number of occasions on which they speak.

Minutes Minutes should record:

(a) the date on which the meeting was held
(b) those present
(c) separate, numbered minutes of the matters discussed recording briefly:
    (i) subject
    (ii) resolution with action to be taken
    (iii) names of person/s to take action
(d) date, time and place of next meeting
(e) distribution list.

Implementing decisions taken This is the third phase of the meeting's progress and in many respects it is the most important because it should produce results. Up to this point ideas have been communicated, proposals generated and decisions made, but it is implementation of the decisions which is the main aim of most meetings. The chair is responsible for implementation but it is frequently the secretary who has the authority to ensure that the persons required to take action are aware of what is expected of them and that they actually do it.

Item 3 on the standard agenda is 'matters arising' (from the minutes of the previous meeting) and a task of the group when considering this item is to review the results of any action taken since the previous meeting.

Sub-committees and working parties These provide the means of implementing decisions of the main 'committee'. If, for example, the management group decides to investigate the possibility of installing an incentive payments scheme for site staff, they might ask a few members of their group to investigate ways and means and report back to the management group. The subcommittee or working party (the terms are more or less synonymous) normally has the authority to co-opt to their subcommittee persons who could assist them with their task.

Such sub-committees are usually ad hoc – set up for a particular purpose – and they are disbanded when their task has been completed.

Advantages of meetings As reference has been made above to their disadvantages it is appropriate to conclude by discussing their advantages. These are:

(a) for meetings of the information and/or instruction-giving type, the advantage of a meeting for the transmitter is that he is able to see and hear the response which his audience makes. This response is spontaneous and the transmitter can, then, if he wishes, adjust his message to match the response. Furthermore, the audience may find a personal, oral transmission more acceptable than the black-and-white message of a memorandum. The interchange of ideas which follows the initial transmission may more than justify the higher cost of a meeting than would be incurred with a memorandum
(b) for generating and interchanging ideas, formulating decisions, marketing and selling, social and staff participation purposes a meeting is almost invariably a suitable medium because it takes into account the human factor. Without an appreciation of the importance of that factor the manager is unlikely to get the best from his collegues.

Sub-contract administration

Appointment and placement of sub-contracts are discussed in section 1.4. This section provides an introduction to administration of sub-contract works during the course of the works from the standpoint of the contractor.

JCT 80 clause 35 contains the *obligations of the employer and contractor* with regard to nomination, renomination, payment and some aspects of extension of the period for completion of nominated sub-contract works.

The obligations of contractor and sub-contractor are contained in the contracts which they enter into which are referred to as sub-contracts.

The sub-contract conditions with which this section is concerned are principally:

- nominated sub-contract conditions NSC/4
- domestic sub-contract conditions DOM/1.

NSC/4 and DOM/1 conditions have much in common but the conditions regarding nomination, renomination, payment and extension contained in JCT 80 clause 35 *do not* apply to domestic sub-contractors.

For virtually all practical purposes the contractor provides the point of contact between sub-contractors and architect. All architect's instructions should be given to the contractor. Those which refer to work to be carried out by nominated sub-contractors should be passed, by the contractor, to the sub-contractor. The contractor should ensure that he is the only point of contact between architect and nominated and domestic sub-contractors.

It is generally the contractor who is in contract with domestic and nominated sub-contractors but exceptionally the employer may enter into an agreement with a nominated sub-contractor.

Reference was made in section 1.4 to the need for the contractor to obtain the consent of the architect before he appoints domestic sub-contractors. Such sub-contractors are to all intents and purposes regarded by the architect as part of the contractor's workforce and not 'recognised' by the architect as separate firms.

The nominated sub-contractor's contractual obligations are similar to those of the main contractor, namely to carry out and complete the sub-contract works in compliance with the sub-contract documents with materials and workmanship of the quality specified and in conformity with all reasonable directions and requirements of the contractor. The quality and standards of materials and workmanship are matters for the opinion of the architect and must be to his reasonable satisfaction (NSC 4 cl 4.1.1).

The directions given to and requirements made of the sub-contractor by the contractor must be reasonable.

The sub-contractor must keep a competent person-in-charge upon the sub-contract works and 'forthwith comply' with architect's instructions (NSC/4 cl 4).

The sub-contractor is obliged to comply with all the provisions of the main contract so far as they relate and apply to the sub-contract works (NSC/4 cl 5).

The contractor's obligations regarding architect's instructions and directions are contained in NSC/4 clause 4.2; the contractor is obliged:

'forthwith to issue to the Sub-Contractor any written instruction of the Architect issued under the Main Contract affecting the Sub-Contract Works (including the ordering of any variation therein); and may issue any reasonable direction in writing to the Sub-Contractor in regard to the Sub-Contract Works'.

Any written instruction of the architect or direction of the contractor given to the sub-contractor's person-in-charge is deemed to have been issued to the sub-contractor.

Failure by the sub-contractor to comply with a direction from the contractor entitles the contractor, subject to the conditions contained in NSC/4 clause 4.5, to employ 'other persons to comply with such direction' and deduct the costs from monies due to the sub-contractor.

Non-completion, extension of time, loss and expense arising from matters materially affecting the regular progress of the works, including the *discovery of antiquities*, are matters of great importance to the contractor. They are matters which may lead to claims from the employer or the contractor and they are the subject of section 2 of this text.

Set-off and adjudication Disputes are discussed in section 4. NSC/4 and DOM/1 contracts contain arbitration clauses.

In addition to arbitration clauses the sub-contract conditions provide for the contractor to 'set-off' (to deduct) the sum he believes to be the value of his claim from any sum which may be due to the sub-contractor for work which he has carried out.

The conditions also provide a safeguard for the sub-contractor by providing for the appointment of an adjudicator to decide the course of action to be taken regarding any amount which the contractor may be disposed to deduct.

Clauses 23 and 24 in both NSC/4 and DOM forms are concerned with the contractor's right to set-off and the action the sub-contractor should take if he disagrees with the set-off. The clause numbers used below refer to NSC/4 conditions.

Set-off may be made with the agreement of both parties. The contractor may also deduct amounts *not* agreed but clause 23.2 imposes three restrictions:

– that no set-off relating to any delay in completion should be made unless the certificate of the architect of failure of the sub-contractor to complete on time has been issued to the contractor with a duplicate copy to the sub-contractor
– the amount of the set-off has been quantified in detail and with reasonable accuracy by the contractor
– the contractor has given to the sub-contractor notice in writing specifying his intention to set-off and the grounds on which the set-off is claimed. The timing of the notice is stated in clause 23.2.

If the sub-contractor disagrees with the set-off he may, within 14 days of receipt of the contractor's notice, send to him by registered post or recorded delivery a detailed and reasonably accurate written statement setting out reasons for his disagreement and particulars of any counterclaim.

At the same time the sub-contractor must give notice of arbitration to the contractor and request action by the adjudicator; sending him copies of his statement and of the notice of the contractor.

The name of the adjudicator is inserted in the appendix to NSC/4 when the parties enter into contract so action may be taken without delay. Should the adjudicator be unable or unwilling to act he should name another person to act in his place.

The contractor has 14 days *from date of receipt* of the sub-contractor's statement in which to send to the adjudicator his response to the sub-contractor's statement.

The contractor may be well advised to inform the adjudicator of the date of receipt of the sub-contractor's statement because in the absence of a statement from the contractor the adjudicator makes his decision.

The adjudicator's action is stated in clause 24.3

Within 7 days of receipt of a written statement by the contractor, or on the expiry of the time limit to the contractor, the adjudicator, without further written statements and without a hearing is obliged to decide how the amount in dispute is to be dealt with. The options open to him are to instruct that:

– the amount is to be retained by the contractor, or
– pending arbitration the amount is to be deposited by the contractor with the trustee-stakeholder, or
– it is to be paid by the contractor to the sub-contractor

or any combination of the above courses.

The adjudicator is not required to give reasons for his decision and his decision is binding upon the contractor.

Clause 21 in NSC/4 and in DOM/1 refer to interim valuations, certificates and payments for sub-contractors. Reference, below, is to NSC/4 but the clauses are similar in many respects.

The contractor is obliged to pass on to the architect any application made by a nominated sub-contractor.

Clause 21.3.1.1. requires the contractor, within 17 days of the date of an interim certificate, to notify to the sub-contractor the amount included in the certificate for him and pay the sub-contractor within 17 days. The sub-contractor is obliged to supply the contractor with written proof of payment to enable the contractor to provide the architect with the 'reasonable proof' required by JCT 80 clause 35.13.3.

Clause 21.4 provides a format for a statement of account for the payment of sub-contractors. It makes provision for payment for work executed, materials and goods not incorporated in the works, deductions for retention and contractor's cash discount.

The sub-contractor may achieve practical competion in advance of the contractor. Obvious examples are piling and ground works contractors.

NSC/4 clause 14.1 requires the contractor, if the sub-contractor notifies him in writing of the date when in the opinion of the sub-contractor the sub-contract works have reached practical completion, to immediately pass to the architect any such notification together with his observations. A copy of the observations must also immediately be sent by the contractor to the sub-contractor.

It rests with the architect to name the date in the certificate of practical completion (NSC/4 cl 14.2).

The sub-contractor's liability for defects under NSC/4 is similar to that of the contractor under JCT 80 to which reference is made in section 1.8, above.

The sub-contractor is obliged to make good at his own cost all defects etc due to materials or workmanship not in accordance with the sub-contract (NSC/4 cl 14.3).

Clause 14.4 provides for making good defects to be undertaken by both contractor and sub-contractor if the architect so instructs.

Domestic sub-contractors have similar conditions under DOM/1 but the contractor directs the making good to be done. DOM/1 clause 14.4 provides that where the architect instructs that making good defects shall not be entirely at the contractor's own cost, the contractor 'shall grant a corresponding benefit to the Sub-Contractor'.

Sub-contractor's final certificate and final sub-contract sum NSC/4 clause 21.10 provides a useful checklist and format for a final statement of the sub-contract sum. Clause 21.11 is concerned with computation of the ascertained final sub-contract sum.

The differences between clauses 21 in NSC/4 and DOM/1 with reference to the final account are greater than those between the other clauses.

The sub-contractor is required to provide all documents necessary for the purpose of the adjustment of the sub-contract sum or the ascertained final sub-contract sum, as the case may be.

Section 1D
PRACTICAL COMPLETION OF THE WORKS

1.9 Practical completion and final payment

Practical completion checklist The checklist for use by project managers for the local authority client, to which reference has been made above, includes under the heading 'Practical Completion' items such as monitoring arrangements for services commissioning, consultations with users (residents) regarding their satisfaction, ensuring their complaints are rectified and providing them with adequate information for operation of specialised plant and equipment and providing demonstrations, setting quality standards for finished work, monitoring 'snagging' undertaken by consultants and following up at later date, obtaining feedback on defects prior to practical completion and making final inspection prior to hand-over, ensuring the site has been cleared, sanctioning issue of certificate of partial possession and/or practical completion, sanctioning extension of time (as appropriate) or deducting liquidated damages or sanctioning part release of retention monies, initiating procedures for consultant to rectify urgent defects during the defects liability period and setting date for final defects inspection at completion of defects liability period. The list also contains items regarding initiating official opening or similar ceremonies and arranging publicity and public relations events.

The checklist provides an effective framework for the contractor's procedures as the project draws to a close. On traditional contracts he is obliged to take action before, for example, snagging can be undertaken or before the certificate of practical completion may be issued.

If the contractor is involved in a design-and-build or other contractual arrangements, his procedures will be very similar to those which give rise to the checklist.

The culmination of all this activity is agreement of the final account and issue of the final certificate which terminates the project.

The procedures regarding the final account and final certificate are, for JCT 80, determined by the conditions contained in clause 30.

Practical completion For the contractor the term 'practical completion' has a particular significance because it marks the 'beginning of the end' of the project.

Practical completion is marked by the issue of an architect's certificate when, in his opinion, 'Practical Completion of the Works is achieved' (cl 17.1).

The Defects Liability Period commences with the certificate of practical completion. As the title indicates, this is the period, at the end of which, the contractor's liability for defects ends. Six months is the period suggested in the appropriate section of the appendix 'if none other stated' by the employer and contractor.

Clause 17.2 states the nature of 'defects' and the procedures to be followed for making good:

'Any defects, shrinkages or other faults which shall appear within the Defects Liability Period and which are due to materials or workmanship not in accordance with this Contract or to frost occurring before Practical Completion of the Works, shall be specified by the Architect in a schedule of defects which he shall deliver to the Contractor as an instruction of the Architect not later than 14 days after the expiration of the said Defects Liability Period and within a reasonable time after receipt of such a schedule the defects, shrinkages, and other faults therein specified shall be made good by the Contractor and (unless the Architect shall otherwise instruct, in which case the Contract sum shall be adjusted accordingly) entirely at his own cost.' (Cl 17.2.)

It is not necessary for the architect to wait until the end of the defects liability period before he issues instructions requiring defects etc to be made good by the contractor at his own cost. No such instructions may be issued after delivery of a schedule of defects or after 14 days from the expiration of the defects liability period (cl 17.3).

The architect issues a 'Certificate of Completion of Making Good Defects' when in his opinion the defects have been made good (cl 17.4).

The contractor is not required to make good at his own cost any damage by frost which may appear after practical completion unless the architect certifies that such damage is due to injury which took place before practical completion (cl 17.5).

The issue of certificates of practical completion and of completion of making good defects are important to the contractor because of their effect on 'retention'.

Retention is an amount which may be deducted and retained by the employer (cl 30.2). The 'retention percentage' is stated as 5% unless a lower rate has been agreed between the parties and specified in the appendix as the retention percentage (cl 30.4.1). Footnote 5 to clause 30.4.1 states that where the employer at the tender stage estimates the contract sum to be £500,000 or over, the retention percentage should not be more that 3%.

The retention percentage is deducted and retained by the employer from the total amount of work that has not reached practical completion. The rules for ascertainment and on treatment of retention are contained in clauses 30.4 and 30.5.

When the architect issues the certificate of practical completion the percentage deductible is one half of the retention percentage (cl 30.4.1).

The remaining retention money is released after the expiration of the defects liability period named in the appendix or after the issue of the certificate of completion of making good defects, whichever is the later (cl 30.4).

Release of retention may, in an industry such as the construction industry which has small profit margins, be a significant factor affecting the contractor's cash flow.

Partial possession by employer There is provision in clause 18 for the employer before the date of issue by the architect of the certificate of practical completion, to take possession 'of any part or parts of the Works', should he so 'wish' provided the contractor consents. The contractor should not withhold his consent unreasonably.

There are various reasons why the employer should wish to take 'partial possession'. By doing so he may, for example, be enabled to let or sell the offices, shops, flats, etc, and so obtain a return on his investment.

There are advantages and disadvantages to the employer taking 'partial possession', for the contractor. Advantages are that when the employer takes partial possession, practical possession of possessed parts is deemed to have occurred and the retention percentage is reduced by one half, as indicated above. The certificate of completion should follow six months later – or whatever is the agreed defects liability period.

Further advantages include the employer taking responsibility for insurance when he takes possession and the contractor receiving alleviation from liquidated damages for that part of the building which has been possessed (cls 18.1.3 and 19.1.4).

Disadvantages are that the contractor retains liability for that part of the building which the employer has not possessed and should, for example, a child from a possessed house be injured by a fall down an open trench – building sites having an irresistible fascination for children – the contractor may well be liable for damages. It may be necessary for additional hoardings or similar protection to be erected.

The contractor may, too, find it necessary to commission building services, connect up drains, etc, in advance of the regular progress of his work. He should seek instructions from the architect before undertaking works which will involve him in additional cost. Such works should then be treated as variations.

The treatment of defects in the possessed part during the defect liability period is the same as that described above for the building as a whole.

Final account The latter part of clause 30 (cls 30.6 to 30.10) describes the method to be used in adjustment of the contract sum, issue of the final certificate or the effect of certificates. Proforma are available which facilitate preparation of a final account.

The contractor's financial management procedures for projects should include final accounting systems which use as models the rules for the ascertainment of amounts due and treatment of retention, etc, described in the conditions of contract. The contractor's accounting systems should be used as a 'shadow' of the quantity surveyor's final account.

In the event of delay in preparation of the final account by the quantity surveyor the contractor's shadow may provide a useful negotiation 'tool', especially when financing charges are high. The contractor may gain tactical advantage by declaring his expectation of the final account amount to the quantity surveyor (and perhaps the employer), as soon after, (if not before), the project is complete. This enables the employer to be made aware of the contractor's expectation and to make accounting provision for, for him, the 'worst case'. The contractor's negotiating position is stronger if such provision has been made by the employer. Negotiation regarding financing charges, to which reference is made in section 2, is facilitated if the contractor is able to demonstrate to the quantity surveyor that he, the contractor, has been held out of funds by underpayment during the progress of the works.

Effect of final certificate Unless arbitration or other proceedings are involved and provided fraud is not involved the effect of the final certificate is that it provides conclusive evidence that:

- effect has been given to the terms of the contract
- extensions of time, if any, have been given
- reimbursement of direct loss and/or expense, if any, has been fully settled.

Should arbitration or other proceedings be involved the final certificate is subject to the arbitration award.

The effect of the final certificate is important to the contractor because for practical purposes it brings the contract (and the project) to an end.

The contractor's administrative procedures should be directed towards expediting production of the final account and remedying defects to ensure issue of the final certificate as promptly as possible. The procedures should include monitoring the quantity surveyor's progress with the final account.

1.10 Alternative methods of terminating the contract

Amicable agreement of the amount due to the contractor is by far the most satisfactory conclusion of the contract.

Regrettably, for various reasons to which reference is made below, the optimum conclusion is not always practicable. JCT 80 provides for alternative conclusions. These include:

- determination by employer
- determination by contractor
- determination by employer or contractor
- outbreak of hostilities.

Determination by employer The contractor's employment may be determined, (brought to an end), for default in four respects or in the event of him becoming bankrupt or acting corruptly. The defaults are:

- without reasonable cause he suspends the carrying out of the works before completion
- he fails to proceed regularly and diligently with the works
- he refuses or persistently neglects to comply with a written notice from the architect requiring him to remove defective work, or
- he fails to comply with the provisions of clause 19 (which refers to assignment and sub-contracting).

If the contractor defaults in one or more of the above respects the architect may give him a notice specifying the default. If the contractor continues the default for 14 days after receipt of the notice (or repeats it) the employer may within 10 days give notice determining the employment of the contractor.

The notice must be given by registered post or recorded delivery and must not be given unreasonably or vexatiously (cl 27.1).

The means by which notice must be given is very specific.

The term 'bankruptcy' is used to include winding up, having a receiver appointed, etc. In the event of the contractor becoming bankrupt his employment is automatically determined but it may be reinstated by agreement between the employer, the contractor and his trustee in bankruptcy, liquidator, etc, (cl 27.2).

The employer is entitled to determine the employment of the contractor if he offers bribes or rewards (cl 27.3).

The rights and duties of the employer and contractor are described in detail in clause 27.4. The employer may employ and pay other persons to carry out and complete the works.

The 'other persons' may use the contractor's plant, equipment and materials on site. The contractor is obliged to assign to the employer the benefit of any agreement he may have for the supply of materials or work.

When instructed to do so by the architect the contractor is obliged to remove his plant and equipment, etc, from site. If he does not remove it within a reasonable time the employer may sell it and hold the proceeds to the contractor's credit.

The architect is obliged to certify the amount of expenses incurred by the employer and the amount of any loss and/or damage caused to the employer by the determination. When the amount has been determined the difference is treated as a debt payable to the employer by the contractor or as a debt payable to the contractor by the employer, as the case may be.

Determination by contractor The contractor may determine the contracts if:

- the employer does not pay the amount due to the contractor on a certificate, or
- he interferes with or obstructs the issue of a certificate, or
- he suspends the works for one or more of the following reasons:
    - an architect's instruction other than one caused by default of the contractor or someone for whom the contractor is responsible
    - the contractor not having received in due time necessary instructions, drawings, etc, from the architect
    - delay in executing work by persons engaged by the employer (as referred to in clause 28, above)
    - the opening up for inspection of work covered up, etc (as referred to in clause 8)
    - failure of the employer to give ingress to or egress from the site of the works, etc, (cl 28.1.3).

The employer's bankruptcy, etc, may be a reason for the contractor determining the contract.

The contractor may give notice by registered post or by recorded delivery to the employer or architect. The notice may not be given unreasonably or vexatiously (cl 28.1.4).

The latter part of clause 28.2 is concerned with the rights and duties of the employer and contractor.

The contractor is obliged to take steps to make safe the site, temporary buildings, etc, and remove his temporary buildings, plant, etc (cl 28.2.1).

The employer is obliged to pay the contractor and to inform him of the 'part or parts of the amount...attributable to any Nominated Sub-Contractor' (cl 28.2.2).

Determination by employer or contractor
As the heading suggests, clause 23A is concerned with the grounds for determination of employment of the contractor by mutual agreement or for one or more of the following reasons:

- force majeure
- loss or damage to the works by one of the specified perils
- civil commotion.

Either party may give notice to the other by registered post or recorded delivery.

The contractor is not entitled to give notice if the loss or damage caused by the specified perils was caused by his negligence or by the negligence of someone employed by him (cl 28A.2).

The contractor is well advised to seek expert advice in the event of determination by either employer or contractor. A small number of quantity surveying firms have developed a special expertise regarding termination of construction contracts.

Outbreak of hostilities and war damage (cls 32 and 33) In the event of an outbreak of hostilities on a scale involving general mobilisation either party may give written notice to the other determining the employment of the contractor. The notice may not be given until 28 days after mobilisation or after practical completion of the works unless the works have sustained war damage.

The architect is obliged to give the contractor instructions regarding protecting or continuing the works and the contractor is obliged to comply with the instructions. If, for reasons beyond his control, the contractor is unable to complete the works he may abandon them.

14 days from the date of notice of deter-

mination or the works abandoned the works are valued and the contractor paid.

Clause 33 sets out procedures for actions to be taken in respect of war damage to the works and for payment for works carried out.

When the original forms of clauses 32 and 33 were published the prospect of 'hostilities' was of instant Armageddon and the conditions appeared irrelevant. Changing international political circumstances now make them appear less critical.

Section 1E
CONSTRUCTION PROJECT COST MONITORING AND CONTROL

Section 1.1 discusses cost planning during stages A to F in the RIBA plan of work; the stages during which the project moves from inception to the point where the design has been determined and the cost may be estimated within fine limits.

1.11 Elemental cost plan as basis for expenditure monitoring

The elemental cost plan is used as the basis for cost estimating during the design stages and it may be used by the client and by the contractor as a framework for monitoring and controlling expenditure by the client.

The cost plan may also be used by the contractor for monitoring and controlling income but elemental cost plans are, by definition, concerned with the cost of *elements* of a building rather than with the cost of *operations* or *activities*. Elemental cost plans are, therefore, of limited use for planning, monitoring and controlling *production* cost.

Thus it follows that elemental costs may be of greater use to the client than to the contractor.

Nevertheless, elemental costs may be available earlier in the project's life than operational costs and so provide both client and contractor with information for initial corporate and project cashflow projection purposes. Two specifications for controls are that they must be timely and congruent from which it follows that imprecise cost information when required is of greater importance than precise information when it is too late to assist in the decision making or control process.

The following example demonstrates the application of control principles using *elemental* cost data. *Production* (or activity) cost data may be similarly applied.

EXAMPLE
The elemental cost plan, figure 1.27, for a private swimming pool, Oxfordshire, (*Building*, April 1989) is used.

For demonstration purposes only *group* costs have been used. More accurate results should be achieved using detailed elemental costs.

Expenditure/Income planning Steps are as follows:
1 The costs are allocated to a 'time-frame' using a network, precedence diagram, bar chart or other production planning method. Figure 1.28 illustrates a network which sets out a logical construction sequence for the project incorporating the group cost items.
2 The contract period for the swimming pool project was 35 weeks and this time-span has been used for the example. Each of the elements is allocated a period of time for its execution. The group element cost is indicated under each element in the network.
  For purposes of the example an approximation of time for each element is calculated on the basis of value using standard production planning methods. The planned duration of each element, in weeks, is shown under the element in the network.
3 Preliminaries are allocated to the whole duration of the project's construction. In the absence of more detailed data a sum of £3000.00 is allocated for 'setting up' the site and £1000.00 for clearing away. The balance is allocated over the

ELEMENTAL COST ANALYSIS

Analysis of contract sum	Preliminaries shown separately Costs at tender date		Preliminaries apportioned among elements		Costs at 2nd quarter 1989
	Cost/element £	%	Cost/element £	Cost/gross floor area £/m²	Cost/gross floor area £/m²
Substructure	46 614	**7.59**	49 539	124.78	161.19
Superstructure					
Frame	163 528	**26.64**	173 788	437.75	565.50
Upper floors					
Roof	28 790	**4.69**	30 596	77.07	99.56
Stairs	—	—	—	—	—
External walls	6 244	**1.02**	6 646	16.74	21.63
Windows and external doors	113 481	**18.49**	120 601	303.78	392.43
Internal walls and partitions	611	**0.11**	649	1.63	2.11
Internal doors	1 545	**0.25**	1 642	4.13	5.34
Group element total	314 209	**51.20**	333 922	841.10	1086.57
Internal finishes					
Wall finishes	7 955	**1.30**	8 454	21.29	27.50
Floor finishes	26 030	**4.24**	27 663	69.68	90.01
Ceiling finishes	1 314	**0.21**	1 396	3.51	4.53
Group element total	35 299	**5.75**	37 513	94.48	122.04
Fittings and furnishings	4 759	**0.76**	5 058	12.74	16.46
Services					
Sanitary appliances					
Services equipment					
Disposal installations	—	—	—	—	—
Water installations					
Heat source					
Space heating and air treatment					
Ventilating system	147 249	**23.99**	156 487	394.17	509.20
Electrical installations					
Gas installations					
Lift and conveyor installations					
Protective installations					
Communication installations					
Special installations					
Builder's work in connection with services					
Builder's profit and attendance on services					
Group element total	147 249	**23.99**	156 487	394.17	509.20
Sub-total excluding external works and contingencies	548 130	**89.29**	582 519	1467.27	1895.46
External works					
Site works	6 226	**1.02**	6 617	16.67	21.53
Drainage	10 264	**1.67**	10 908	27.48	35.50
External services	7 868	**1.28**	8 362	21.06	27.21
Minor building work	5 146	**0.84**	5 469	13.78	17.80
Group element total	29 504	**4.81**	31 355	78.99	102.40
Preliminaries	36 240	**5.90**	—	—	—
Totals excluding contingencies	613 874	**100.00**	613 874	1546.26	1997.86

1.27 *Elemental cost plan for swimming pool Building,* 21 April 1989

duration of the project on a monthly basis.

EXAMPLE

£36,000.00 (prelims total) − £3,000.00 (set-up) + £1,000.00 (clearaway)

= £32,000.00 = $\dfrac{£4,000.00/\text{month}}{8 \text{ months}}$

4 A bar chart, figure 1.29, is prepared assuming monthly payments. The calendar would be used to determine the duration of each month. 4 week or 5 week provisions are used in the example.

5 The values of expenditure/income for each group element each month are

ELEMENTAL COST PLAN AS BASIS FOR EXPENDITURE MONITORING

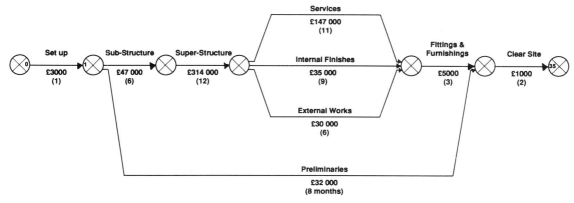

1.28 Costed network for swimming pool

OPERATION	Value £	1 2 3 4	5 6 7 8 9	10 11 12 13	14 15 16 17 18	19 20 21 22	23 24 25 26 27	28 29 30 31	32 33 34 35
Sub - Structure (6)	47 000	23 500	23 500						
Super - Structure (12)	314 000		52 333	104 666	130 833	26 167			
Internal Finishes (9)	35 000					11 667	19 444	3 888	
Fittings & Furnishings (3)	5 000							1 667	3 333
Services (11)	147 000					40 094	66 818	40 091	
External Works (6)	30 000					15 000	15 000		
Preliminaries Set up (1) 3000 Hire (8 months) 4000 Clear up 1000	36 000	set up 3 000 hire 4 000	4 000	4 000	4 000	4 000	4 000	4 000	clear up 1 000 4 000
Monthly			30 500	79 833	108 666	134 833	96 928	105 262	49 646	8 333
Cumulative	614 000	30 500	110 333	218 999	353 832	450 760	556 022	605 668	614 007

1.29 Costed bar chart for swimming pool

calculated, pro-rata the duration of the element.

EXAMPLE
The cost of the sub-structure in month 1
= £23, 500.00 calculated:

$$\frac{£47,000.00 \text{ (total cost of element)} \times 3 \text{ (week occupied on that element during month 1)}}{6 \text{ (weeks duration of element)}}$$

The total of the values for the parts of the elements completed during each month is shown at the foot of the bar chart as monthly and as cumulative values.

Read as a whole, the 'costed' bar chart provides a valid expenditure/income plan which both client and contractor may use for their respective purposes at project and at corporate levels.

In practice allowance must be made for the effect of retention monies and the delay between time of valuing works in progress and time of payment. Such allowances have been ignored in the example.

Monitoring and controlling expenditure/income

6   The costed bar chart, figure 1.29, in itself provides a basis for cost monitoring and controlling but extrapolation to determine trends is facilitated if the bar chart is converted to graphical form as shown in figure 1.30.

    As payments are made at regular intervals, the stepped profile is the more accurate form, but the 'S' curve more readily demonstrates the expenditure/income pattern.

7   To monitor 'plan' in relation to 'achievement' the actual value of work carried out each month should be plotted on the graph and compared with the plan.

8   Forecasts of future expenditure/income may be made by extrapolating the achievement line.

9   The value of work still to be completed is frequently an accurate guide to the actual progress of the works. A comparison of planned with achieved expenditure/income may, therefore, provide a reasonably realistic guide as to the (revised) anticipated completion date and expenditure/income flow.

Integrated cost management Construction project cost monitoring and control should be part of a system which integrates all projects (contracts) within the corporate financial management system.

Budget based corporate financial management is outside the scope of this text but in order to indicate the context in which the contract is administered it is appropriate to outline the principles of corporate cost management.

Cost control may be defined as the regular and frequent comparison of actual expenditure with predetermined standards in the form of budgets so that undesirable trends away from the standard may be detected and corrected at an early stage.

A budget may be defined as a pre-determined plan, expressed in financial terms, covering all phases of a business. A budget should state specifically what is expected of the business in terms of profitability and how the plan is to be achieved. It should be a well conceived plan which ensures the desired level of profitability allied to all the resources available.

Corporate organisation structure In a typical medium-to-large construction 'group' (of companies) the corporate organisation structure might comprise, 'at the top' a *group board* directing the activities of the *divisions* which comprise the group. The activities of the various divisions might include general contracts, speculative housing, commercial developments, plant hire, timber products, etc. Each division would undertake *projects* which would be the sources of the group's revenue.

The group board would represent level 1, the divisions level 2 and the projects level 3.

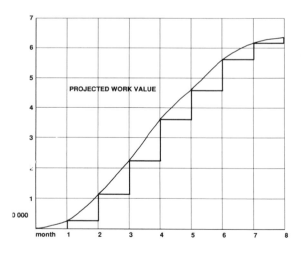

1.30 *Expenditure forecast for swimming pool*

Budgetary control principles and benefits
The fundamental principles of budgetary control are that the budget for the superior level, level 1, should be based on and prepared by the subordinate level, level 2 (and so on) and that the budget is prepared in a form that enables it to be used as part of the corporate integrated budgetary system and provides the basis for control at all levels of the organisation.

The principal benefits of a formalised budgetary control system may be summarised as:

- setting out in financial terms the objectives of the organisation
- providing yardsticks (standards) by which to measure efficiency at all levels in the organisation
- demonstrating the extent by which results vary from anticipation (variances)
- providing a guide for corrective action
- making possible control 'at the top' (level of management) with delegated responsibility.

An integrated financial management system provides the basis for control at all levels of the organisation.

Budgetary responsibility assumes that all sections, divisions, projects, etc, of the group or company have *controllers* each of whom is responsible for the cost and/or profit of his 'section'. It assumes a number of profit or cost centres which are, typically, arranged in levels which correspond with the organisation structure of the enterprise.

Construction project profit centres and budgets The nature and form of profit centres and budgets for cost monitoring and control at construction project level are determined, principally, by the method used to estimate for the work and the practicality of cost allocation for purposes of reconciling the cost expended with the value of the work as executed.

The preparation of the estimate is discussed in section 1.3. Construction work is customarily quantified and 'billed' in *finished work units* (x m^3 of concrete in beds, y m^2 of block work in walls, etc) rather than in activity or operational form.

(The terms 'activity' and 'operation' have a similar meaning in the context of construction planning and control. The term 'operation' is used below, in this section.)

Whilst quantifying and billing work in operational form has advantages for cost monitoring and controlling, the preparation of the bills and estimating take longer. As the success rate for tendering is probably between 10% and 20%, time saving at the estimating stage is important. Attempts to promote operational bills have been strongly resisted by estimators.[7]

For monitoring and controlling the cost of work in progress, work billed in operational form is preferable because the work itself is carried out in a number of operations, not in finished work units.

One expedient is to convert the cost data from *finished work unit* to *operational* form. Thus, '3,000 m^2 of block work in walls' (the finished work form) may be allocated to several operations in different part of the building carried out at various times in the construction programme.

Figures 1.27 and 1.29 in the example show elemental costs allocated to operations. The costs allocated to each operation may be further separated into labour, plant and materials.

Computer applications to budgetary management[8] The alternative to activity or elemental based models for forecasting construction project expenditure/income patterns is the mathematical-based model. The mathematical model uses a mathematical expression to generate different forecasts by altering the value of its parameters.

There are certain characteristics which distinguish between a mathematical model and a non-mathematical model. Normally, a mathematical model has the following properties:

- it is generally cheap and easy to use
- it often needs little experience and knowledge about the project
- it requires only a short period for its operation

- it can be used at an early stage in the project's life
- it tends to alienate the user (connotations of the 'black box')
- it is relatively inaccurate.

A non-mathematical model may:

- be expensive and complicated
- require a great deal of information and know-how about the project, hence it may be laborious
- involve the user in the process of generation of forecast
- be relatively accurate.

The model discussed below combines the most appropriate features of the mathematical model with those of the non-mathematical model, hence, the name TASC (The Advanced S Curve).

The model uses a new mathematical expression and has a comprehensive database consisting of a series of predictive models developed by analysing a large quantity of data.

The forecasting logic It has been statistically validated that the 'shape' of the expenditure/income profile embodies characteristics associated with the physical properties of the project. It has been concluded that as well as the general properties associated with the growth curves, the expenditure/income pattern possesses certain characteristics which distinguish one project from another.

These properties are referred to as dependent-variables. They are listed and briefly described below:

1 The position of the peak point on both the time axis (Xp) and the value axis (Yp). The peak point (P) is the point with the highest value. At this point the rate of change of values is Zero.
2 The position, duration and intensity of any distortion to the underlying pattern of expenditure/income. (A distortion also produces a trough.)
3 The measure of the overall rate of growth from the origin to the main peak point.
4 The measure of the overall rate of decay from the main peak to the end of the project.

The above characteristics are highlighted in figure 1.31.

Generation of a forecast In order for the mathematical expression of TASC to generate a forecast, the values relating to the dependent-variables must be known. See figure 1.32.

Modes of operation The above dependent variables can be estimated either manually by the user or automatically by TASC.

Manual mode This mode is intended for both experienced and experimenting users. The latter can obtain a better appreciation of the logic of forecasting and learn about the behaviour of the expenditure/income profile by experimenting with the model. The values of the dependent-variables can be altered and the subsequent changes on the expenditure/income pattern can be observed.

The experienced user can, however,

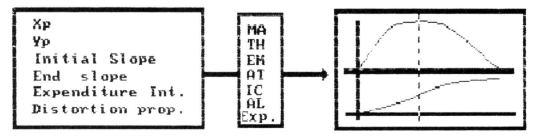

1.31 Properties of TASC, Generation of forecast for given values of dependent variables

ELEMENTAL COST PLAN AS BASIS FOR EXPENDITURE MONITORING

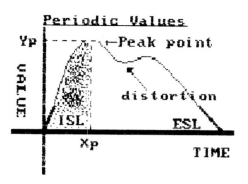

1.32 TASC dependent variables

exploit his experience and knowledge-based centre (brain) to estimate the values of the dependent-variables. There are two options within the manual mode.

The first is the construction of an approximate barchart which can be constructed to encompasses the major elements or centres of cost. A twenty stage smoothing process is carried out by TASC in order to evaluate the values of the dependent-variables.

The second is *direct entry* In this mode the values of the dependent-variables are given by the experienced users.

Automatic mode 87 models have been developed which for given project characteristics estimate the values of the dependent-variables.

In this mode, the user is required to define the characteristics of the project. These characteristics consist of project type, sub-type, operation, scope, form, ground condition, access in horizontal and vertical directions, complexity and any one of the identified twenty three cases which would cause a distortion (eg adverse weather conditions).

In this mode, TASC will also estimate the project total value and duration.

Simulation Once the dependent-variables are estimated, manually or automatically, the mathematical expression generates the forecast by simulating its constituent three modules. The control, distortion and Kurtosis modules are individually generated by means of a newly developed mathematical expression. These modules contribute, independently, towards construction of the overall pattern.

The control module is responsible for simulation of the underlying pattern of expenditure/income as well as positioning the peak point. This module is simulated by the following exponential expression:

$Y = C(e^{ax^b(1-x)^d} - 1)$
C = total project cost
Y = periodic value
X = period/total period
a, b, d are the parameters of the model
where:
$a = d.R/(1-R)$ and
$b = \log(1+Q)/(R(1-R))$

(R) directly controls the time coordinate of the peak point
(Q) directly controls the cost coordinate of the peak point.

The distortion module is shown in figure 1.33. If at any stage of construction, the rate

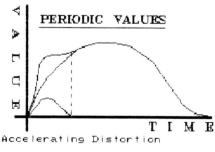
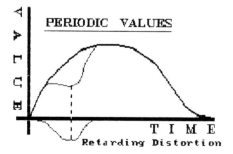

1.33 TASC distortion module

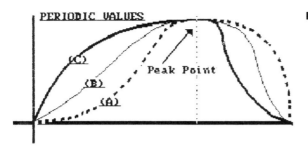

1.34 TASC Kurtosis variables

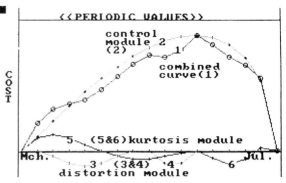

Figure 1.35 Periodic and cumulative values

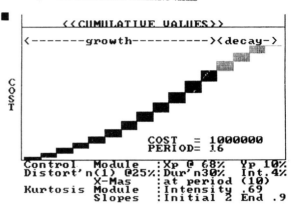

1.35 Periodic and cumulative values

of production/expenditure is retarded or accelerated appreciably, the underlying pattern will be distorted, and secondary peaks and troughs are produced. The distortions have been measured from the data. The accommodation of the distortion module in the overall model is carried out by the following fourth degree polynomial expression which requires information relating to the position, duration (d) and intensity (I) of each distortion.

$f(X) = a1.x^4 + a2.x^3 + a3.x^2 + a4.x + a5$
$f(0) = 0 \quad f(1) = 0 \quad f'(0) = 0 \quad f'(1) = 0$
$a1.K + a2.L + a3.M + a4.N = I$
(I = the intensity of distortion)
where:

$$K = \mathrm{SUM}(x/d)^4 \quad \substack{x=d \\ x=1} \qquad L = \mathrm{SUM}(x/d)^3 \quad \substack{x=d \\ x=1}$$

$$K = \mathrm{SUM}(x/d)^2 \quad \substack{x=d \\ x=1} \qquad N = \mathrm{SUM}(x/d)^1. \quad \substack{x=d \\ x=1}$$

The Kurtosis module is illustrated in figure 1.34. Different periodic patterns of expenditure may have different measures of peakedness. The adjustment for this measure has been facilitated by the Kurtosis module. Simulation of the Kurtosis module is based on the evaluation and classification of the cumulative expenditure from the origin to the peak point (Kurtosis intensity), and the slopes at the very beginning and the very end parts of the periodic expenditure curve (initial slope and end slope). This module is also simulated by means of a fourth degree polynomial expression similar to that used for the distortion module. The only difference is that $f'(0) = S$ where S is the slope of the curve at its origin. The point of origin is the first period for the initial slope and the last period for the end slope.

The model The above modules are combined to construct the overall project expenditure pattern. The simulated pattern satisfies all the requirements which have been identified through the manual mode or the automatic mode. This has been demonstrated in figure 1.35.

TASC software The above model forms the core of the TASC software which is a comprehensive project financial management tool.

TASC generates financial reports for both the client and contractor in both numerical and graphical forms. The reports consist of:

- project income and expenditure flow
- project profit flow
- project retention flow
- project cash flow
  - net cash flow
  - maximum cash flow
  - real cash flow
- growth and decay of cash based on interest charges and gains
- modular break down of financial components
  - control module
  - distortion module
  - Kurtosis module

TASC makes optional allowance for:

- Christmas breaks
- front end loading
- retention
  - % retention application
  - % release on practical completion
  - defect liability period
- periodical interest charge and gain for – and + cash
- flexible period intervals.

Updating/monitoring After the initial forecast the values will be known period by period. These values can be entered and stored into TASC at desired intervals. At each stage a new forecast is generated and the corresponding interest charge and gain are identified. The actual values are then contrasted, both numerically and graphically, with the original forecast and the forecast generated during the previous period.

This section on computer applications is concerned with application of a particular program for a specific purpose but programs are available for the management of other processes such as the reconciliation of cost with value discussed below.

Cost:value reconciliation of work in progress The CIOB's manual, *'Cost:Value Reconciliation'* provides comprehensive principles for the contractor.

The manual adopts the ICA's statement of standard accounting practice (SSAP) No. 9 – stocks and work in progress.

SSAP 9 provides that assessments of work in progress in periodic financial statements should be: 'Cost, plus attributable profits (if any), less foreseeable losses (if any).'

Attributable profits are defined as:

'that part of the total profit currently estimated to arise over the duration of the contract (after allowing for likely increases in costs so far as not recoverable under the terms of the contract) which fairly reflects the profit attributable to that part of the work performed at the accounting date.'

'There can be no attributable profit until the outcome of the contract can be assessed with reasonable certainty.'

Foreseeable losses are defined as: 'losses which are currently estimated to arise over the duration of the contract (after allowing for estimated remedial and maintenance costs, and increases in costs so far as not recoverable under the terms of the contract).'

Profits can only be taken when they are earned and if a contract is in loss in its early stages, profits cannot be drawn ahead.

Variations should be identified and valued as quickly as possible after they occur. The value should be agreed with the quantity surveyor if it is to be taken into account in the reconciliation.

Contractual claims from sub-contractors (and others) which have yet to be resolved and liquidated damages should also be taken into account in the reconciliation.

Barrett explains in detail the operation of cost:value reconciliations which should provide a model for the contractor. The operation includes adjustments to make in regard to delay, allocation of a preliminaries budget to a time scale and provisions to be made in respect of future losses on work still to be carried out, estimated future costs on rectification and guarantee work, foreseen claims and penalties arising from delays in completion or from other causes and non-recoverable increases in costs.

Reconciling the cost:value of operations
The procedures described by Barrett are concerned primarily with determining the value of 'work in progress' for statutory and

corporate accounting purposes. In that respect they provide for the needs of the divisional director and financial manager for the group.

Such reconciliations depend on data provided by sub-contractors, merchants and others. There may be a significant lapse of time between valuation of the work in progress and ascertainment of the adjusted 'prime cost'.

The construction manager or managing surveyor responsible for the contract's financial management may require a means of monitoring cost and progress which may be applied within hours of valuation and which reconciles the cost and value of operations. They require more detailed, smaller 'budgets'.

To this end budgets prepared in a similar manner to those shown in figures 1.29 and 1.30 but based on production costs rather than on elemental costs are more useful.

Such operational *budgets* may be compared with the *value* of work on each operation at the time of the valuation in order to ascertain profit or loss.

Valuations prepared in this way may not be as accurate as those prepared adopting the procedures described by Barrett but they facilitate an earlier means of monitoring than Barrett's and enable the contractor to take prompt action to exercise control and take remedial action in the event of unsatisfactory trends being identified. He is able, too, to identify which of a number of concurrent operations is profitable and which is not.

Recording operational costs Reasons for recording the time spent on and the cost of operations include:

- monitoring and controlling the budgeted with the actual costs of operations
- providing data for valuing variations
- providing data for ascertaining loss and expense caused by disturbance of the regular progress of the work
- providing data for valuing work carried out by operatives being rewarded through productivity based incentive payments so that the work value can be compared with target costs.

Interim financial reports should contain:
- initial tender figures and expected profit
- forecast figures at completion for value and profit
- current payment application by the contractor
- current certified value
- adjustments to and provisions in the certified valuation
- costs to date of the accounting period in question
- cash received to date, retention deducted and certified sums unpaid.

Figure 1.36 shows a proforma which may be used for standard reports.

In order to record operatives' and plant time, it is necessary to identify the operations to be costed and the time spent by operatives and plant on the operations. It is important that the time spent on the operations is recorded as promptly as possible after the event.

The allocation sheet should be completed by the foreman, change-hand or other 'leader' immediately responsible for the team whose names appear on the sheet.

Figure 1.37 provides a matrix which facilitates daily allocation of time spent by labour and plant on operations. A rule is that all time expended must be recorded to an operation (cost centre). Operations should be predetermined. If time has been spent on activities which have not been predetermined, the causes should be identified.

Such causes may include work arising from architect's instructions which comprise variations, work not arising from architect's instructions but which cannot be allocated to an operation, remedial work (which may result from managerial inefficiency or other causes) or causes which initially, at least, are unaccountable. Identifying a cause soon after an event may enable the contractor to take action to avoid loss.

By allocating time spent by plant operations, the contractor is able to ascertain the actual cost of plant on operations for comparison with the estimate. He is also able to ascertain the ratio of working to standing time.

ELEMENTAL COST PLAN AS BASIS FOR EXPENDITURE MONITORING

1.36 Cost:value reconciliation standard reporting form
BARRETT F R, 1981

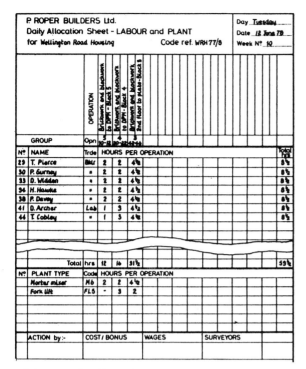

1.37 Daily allocation sheet (for labour and cost allocation)

The allocation sheet is a useful tool for providing the costing data which can be used to provide the standards used for estimating, to which reference is made in section 1.3.

1.12 Case study

Hawthorne Home Development provides a 'role play' scenario intended to develop the participant's appreciation of his/her contractual obligations, rights, etc.

The scenario provides case material for consideration of action which should be taken in given situations to expedite the project and to safeguard the interests of those concerned. The clause/s in the conditions of contract which are relevant to each situation should be identified and stated.

The participant should assume a 'preferred' role. If in that role he is not directly concerned in the situation he should suggest the action which should be taken by those concerned.

Scenario

(a) Project for construction of three storey home for elderly plus ancillary buildings in grounds of former Victorian industrial building which is to be altered and extended to provide part of accommodation.

PROJECT TITLE: HAWTHORNE HOME DEVELOPMENT (HHD)

(b) *Client:* Retirement Homes Limited
(c) *Design by:* Able Design Associates
(d) *Cost control by:* Cost Management Associates
(e) *Construction by:* South London Construction Limited
(f) *Tender date:* 10 February
(g) *Date of contract:* 31 March
(h) *Commencement date:* 14 April
(i) *Contract period:* 70 weeks
(j) *Contract sum:* £1,000,000
(k) *Liquidated damages:* £15,000 per week
(l) *Form of contract:* JCT Standard Form of Building Contract, 1980 edition with quantities, NSC/4 condition for nominated sub-contracts, DOM/1 for domestic contracts.

Situation

Date of contract + 10 days
Architect, on behalf of employer, gives notice that possession of site will be delayed by 3 weeks.

Date of contract + 14 days
Architect issues instruction to contractor that work included in contract bills, for execution by named contractor, O. Riginal, is now to be carried out by another person as instructed later.

CASE STUDY

The following situations arose after the date of possession

Week 1
Bundle of drawings is delivered to contractor's office from architect. Drawings are marked with revision letters from G to J. List of drawings in contract bills refers to revisions B and C. List of drawings in articles of agreement refers to revisions E and F. The drawing numbers of those in bundle and both lists are essentially the same.

Week 2
Bundle of drawings is delivered to contractor's office from consultant services engineer under cover of a compliments slip.

Week 3 (1)
Checking through architect's drawings, the contract manager finds details of dormer windows shown on drawings are of preformed glass fibre 'one-piece' units whereas the contract bills describe dormers as constructed of timber with lead heads and cheeks.

Week 3 (2)
Assistant architect visits site and whilst making inspection tour with site manager tells him to:

- reposition temporary hoarding 300 mm inside back of footpath (which is site boundary)
- increase thickness of concrete foundations by 150 mm
- instruct contractor's men to work half an hour overtime for the time being because job seems to be behind programme.

Week 3 (3)
Architect, during site visit, tells site manager he likes to be able to refer to prices in contract bills so will site manager arrange for priced copy to be kept on site.

Week 4 (1)
Excavation sub-contractor excavating foundations damages 1950 mm oval brick sewer not shown on drawings. Sub-contractor says he is being held up, waterboard engineer says sewer is disused but must be cut back to site boundary by contractor and ends at boundary sealed by waterboard men for which there will be a charge.

Week 4 (2)
An ancillary building is shown on setting-out drawing to be constructed 15 m from existing factory building and parallel with it. Whilst excavation is in progress it becomes apparent that set out as detailed there will be insufficient space between ancillary buildings and home to accommodate covered walkway, service road and flower beds which are shown on another drawing but are not dimensioned on the setting-out drawings.

Week 5
Excavator (on hire) uncovers ancient brickwork and stonework 1 m below surface. A mosaic pavement is exposed. Driver brings find to notice of site manager saying there is nowhere else he can work and shall he phone hirer to move him to his next job. Approximately one third of foundations are still to be excavated.

Week 6
Contractor writes to architect saying he has just been awarded major contract and intends to assign Hawthorne Homes Development.

Week 7
Contractor writes to architect seeking approval of Bloggs as sub-contractor for glazing work. Architect responds that Bloggs is unacceptable because he is understood to refuse employment to coloured persons and to be a member of a political party holding strong racist views.

Week 8
Contractor reviews progress against programme and ascertains events in weeks 4 and 5 appear to have delayed regular progress of works by two weeks and cost records indicate he has been caused loss and expense of approximately £15,000.

Week 11
Architect, during site visit, notices truck parked nearby owned by plastering contractor who 'walked off' another job for which he, the architect, was responsible. His enquiries reveal plastering contractor is working as sub-contractor on HHD.

Week 13
Contractor reviews his records and finds no acknowledgment of his notice of delay and application of direct loss and expense.

Week 15
Local resident enters site during working hours and falls down half-built manhole, injuring himself.

Week 22
Inexplicably, substantial cracks appear in property adjoining.

Week 26
Quantity surveyor visits contractor's store at his yard to inspect joinery which contractor has included in an interim application for payment on account. Storeman waves his arm towards end of shed and says the joinery is 'some of that lot'.

Week 29
Contractor refuses to remove defective brickwork: in his view...'the work is not defective...' he says.

Week 30
Contractor is seen preparing to continue bricklaying, building in the defective work which he was instructed to remove two weeks earlier.

Week 32
Christmas is approaching and van arrives at architect's office with one case of Scotch and six cases of wine. Label in Christmas card from contractor states '...for the Christmas party'.

Week 35
Nominated sub-contractor for plumbing and heating (P&H) work is declared bankrupt. Receiver is appointed and informs contractor of his appointment.

Week 40
Nominated sub-contractor for plumbing and heating work uses blow-lamp to joint his pipes fixed to existing lathe and plastered partition in existing industrial building. During subsequent night fire breaks out causing damage to existing building, new works carried out by contractor, P&H sub-contractor and other sub-contractors. P&H sub-contractor's own lock-up with his materials and plant gutted. Inspection on 'morning-after' establishes fire caused by sparks from blow-lamp igniting shavings in cavity of partition.

Week 44
Contractor has not received interim payment for which he received copy of architect's certificate in week 41. This is not first occasion on which this has happened (employer's computer has been playing up and earlier there was a dispute with computer operators).

Week 45
Architect writes to contractor stating the employer's intention to take possession of refurbished and extended Victorian building in week 48. (The works in these buildings is approaching completion.)

Week 70
Vandals strike over weekend, 80% of glazing broken, doors smashed, plumbing fittings and pipes wrecked. Contractor's first estimate of delay is five weeks and he writes to architect to that effect.

Week 78
Architect issues certificate that practical completion of the works has been achieved. He has extended completion date by four weeks. Employer deducted four weeks liquidated damages from previous interim certificate and architect says he understands employer will retain the liquidated damages previously held despite contractor's objections at time of previous certificate.

1.13 Exercises

1. *Comment on* the relative merits to the employer and the contractor of using the Code of Procedure for Selective Tendering as a basis for obtaining tenders.

2. The selective tender process offers an implied assurance to the client of the level of expertise and the financial adequacy which can be expected of those contractors tendering. Where firm bills of quantities also form part of the contract documentation there exists the further implication of 'certainty of price'.
   Discuss.

3. The use of bills of quantities as a contract document may be considered to be an outmoded practice.
   Discuss this point of view.

4. Most standard forms of contract contain provisions which seek to protect the client's financial position in relation to the contractor's performance.
   Discuss the extent to which the following provisions afford this protection:
   (i) retention
   (ii) liquidated damages
   (iii) performance bonds.

5. In recent years the traditional functions of the quantity surveyor have been subject to change. *Discuss* the nature and reasons for such change and describe the likely role of the quantity surveyor in the future.

6. For building works where time is an important factor, it is common practice for a letter to be issued by an architect expressing the employer's intention to enter into a contract on the basis of the tender submitted. This letter may ask for certain preliminary work to be undertaken or that work be commenced on site prior to contract documents being signed.
   Draft a reply to such a letter, identifying the issues.

7. (a) *Discuss* the various contractual arrangements under which specialists are employed in building contracts
   (b) *Describe* the provisions of the IFC 84 in relation to the employment of a sub-contractor of the client's choice.

8. *Discuss* the value of a pre-tender plan at the estimating stage and during the construction period.

9. Where tenders for construction work are to be based upon bills of quantities prepared by the quantity surveyor, substantial portions of the work are often the subject of PC sums. This work will be let at a later stage to nominated sub-contractors many of whom design their own specialist work.
   Discuss
   (a) the effect this has upon the contractor's estimating team in predicting the cost of the project.
   (b) the advantages and disadvantages to the main contractor of a relatively high level of nomination.

10. (a) *Describe* the roles of the following individuals in the valuation of work and certification of payment, as described in JCT 80, Private with Quantities.
        (i) architect/supervising officer
        (ii) quantity surveyor
        (iii) employer.
    (b) *Explain* how the contractor can assist in the valuation process so as to ensure that his work is fairly valued.

11. The rules for the valuation of variations applying to work carried out under JCT 80 may be considered to be inadequate and to be the primary cause of many disputes.
    Discuss techniques and procedures which may be adopted by site management as an aid to establishing cost and gaining adequate financial reimbursement.

12 The incorporation of bills of quantities as a contract document has certain effects.
   Discuss the rights of the contractor under JCT 80 where:
   (a) the drawings and quantities differ
   (b) rock is unexpectedly found whilst carrying out excavation work
   (c) an architect's instruction is issued which considerably changes the quantity of work in a particular element.
   (Clause numbers are to be quoted where appropriate.)

13 A contract is let under JCT 80. The employer has expressed dissatisfaction with the finish of the concrete floor slab and annoyance with the fact that the contractor is falling behind the construction programme. As a result the employer decides not to pay the last interim certificate recently issued by the architect and informs the contractor of this fact by telephone. The contractor immediately removes all labour and plant resources from site and informs the employer by telephone that resources will not be returned to site until payment is received.
   Advise both parties, quoting clause references where appropriate.

14 A contract let under JCT 80 for the construction of a four storey traditionally built office block has been delayed at an early stage by unexpectedly bad weather and later by the slow performance of the nominated heating sub-contractor. As a result of these delays the contractor anticipates that this contract will make a considerable loss and he is considering abandoning the project before his financial situation worsens.
   Appraise the contractor's contemplated action and discuss the rights he may possess to remedy the situation. (Clause numbers are to be quoted where appropriate.)

15 A building contract made under JCT 80, Private with Quantities, has a single completion date. During the course of the contract, the employer informs the contractor that he wishes to take possession of part of the work which has already been completed.
   Comment upon the rights of the employer in this situation and describe the actions which the contractor should take.

16 A contract of £2.25 m, being carried out under JCT 80, was suspended when the work was approximately 70% complete and the employment of the main contractor determined due to bankruptcy.
   Examine
   (a) the possible contractual issues which may have contributed to the insolvency of the main contractor
   (b) the position of nominated sub-contractors.

17 An architect issues an instruction which substantially increases the amount of excavation in a contract and directs the contractor in his method of excavation.
   (a) *Compare* the architect's authority to issue such instructions under both JCT Edition, Private with Quantities and IFC 84
   (b) *Discuss* the factors to be borne in mind by the contractor and the architect in valuing this varied work.

18 Having regard to the scenario in section 1.12, *explain* the appropriate action to be taken with regard to each of the situations to safeguard the interests of the parties concerned.

19 Insert the earliest and latest completion dates in figure 1.28 and *determine* the critical path/s.

20 *Insert* the monthly cumulative totals in figure 1.29. Is the cumulative total shown against month 8 correct?

21 Assuming the cumulative value of work actually carried out on the swim-

ming pool project during months 1 to 5 was as follows:

Month	Cumulative value
1	£ 27,000
2	85,000
3	170,000
4	300,000
5	400,000

Plot the values on figure 1.30 and *extrapolate* to forecast the anticipated actual construction time. *Calculate* revised expenditure/income sums for the remaining months of the revised construction period.

22 Using more detailed data taken from figure 1.27 *undertake* a more accurate exercise following the steps outlined above.

1.14 Sources and further reading

Relevant documents, forms of contract, etc

JCT Standard forms of building contract
— private with quantities, 1980 edition
— with contractors design, 1981 edition
— intermediate for works of simple content, 1984 edition
— management contract, 1987 edition
— nominated sub-contract NSC/4, 1980 edition
— sub-contract conditions NAM/SC, 1984 edition
— works contract conditions/1 and 2, 1987 edition

BEC sub-contract conditions for use with the domestic sub-contract DOM/1, 1980 edition
JCT Arbitration Rules, 1988

Corporate organization structures, management systems, financial management (corporate and project)

FELLOWS R, LANGFORD D, NEWCOMBE R and URRY S, *Construction management in practice*, Construction Press, 1983
KAST F E and ROSENZWIEG J E, *Organization and Management*, McGraw-Hill, 1985
BURGESS R A and WHITE G, *Building production and project management*, Construction Press, 1979 (which provides an appendix of roles, job titles, tasks of construction personnel)

Cost planning and project appraisal methods

SEELEY I H, *Building Economics*, MacMillan 1987
BCIS (manages the most extensive collection of cost data which is available to subscribers to the services)

Estimating and tendering

PARK W R, see references in 1.15, number 4, 'North American methods and perspective'
CIOB, Code of estimating practice, CIOB (definitive checklist on procedures for traditional and non-traditional works)

Sub-contract management

FRANKS J, *Building Sub-contract Management*, Longman, 1984 (from the sub-contractor's standpoint)
HIBBARD P, Sub-contracts under the JCT Intermediate form, BSP Professional Books, 1987 (detailed study of IFC 34)
FRANKS J, Sub-contract conditions associated with the JCT intermediate form of contract – a view from the sub-contractor, CIOB Technical Information Service 100 and 101

For co-ordination of documentation

CCPI Co-ordinated project information CCPI, 1987
CCPI Common arrangement of work sections for building works, CCPI, 1987
CCPI Production drawings, CCPI, 1987
CCPI Project specification, CCPI, 1987

BARRATT F R, *Cost:value reconciliation*, CIOB 1981 (provides the most comprehensive guide for financial monitoring and control)

1.15 References

[1] RICS Central London Branch report on project management, *Chartered Surveyor*, April 1977.
[2] Junior Organisation Quantity Surveyors

Division, *An introduction to cost planning*, RICS, 1976.
3. FISHER N, *Marketing for the Construction Industry*, Longman, 1986.
4. PARK W R, *Construction Bidding for Profit*, Wiley, Chichester, 1978.
5. The committee comprises members of all disciplines concerned with building procurement.
6. CHAPPELL D, 'Just a minute', *Building*, 26 September, 1980.
7. BRE carried out research into operational bills and operational format bills which proved unpopular with estimators.
8. This section describes the design and operation of TASC, a computer programme used for forecasting construction project expenditure/income patterns, obtainable from FAZTEC Computers, 92 Greatdown Road, London W8 1AP (telephone 081 578 3086).

APPENDIX 1(a)   *Index to procedures manual*

1. **INTRODUCTION**
   General Notes
   Schedule of Company Forms in Use

2. **MATERIALS PURCHASING**
   Requisitions and Purchasing Procedures
   Suppliers Orders
   Nominated Suppliers

3. **SUB CONTRACTORS**
   Enquiries and General Procedures for Placing Orders
   Conditions of Order   Sub Contractor's Site Progress Reports
   Letters of Intent   Pre-Contract Suspense Account

4. **"JCT 1980" FORM**
   Procedures for Placing Nominated and Direct Sub Contractor's Orders

5. **"IFC 84" FORM**
   Procedures for Placing Named and Direct Sub Contractor's Orders

6. **"JCT 1981 WITH CONTRACTOR'S DESIGN" FORM**
   Procedures for Placing all Sub Contractor's Orders and the
   Implications of Design Responsibility

7. **"GC/WORKS/1 – EDITION 2" FORM**
   Procedures for Placing Nominated and Direct Sub Contractor's Orders
   under Government Forms of Contract

8. **SUNDRY FORMS OF CONTRACT**
   JCT Minor Works 1980 Edition
   FAS Form 1975 Edition (1978 Revision)
   Fixed Fee & Prime Cost Form 1967 Edition (1976 & 1987 Revisions)
   Procedures for Sub Contractor's Order under these Sundry Forms
   of Minor Works Contracts

9. **DOMESTIC SUB CONTRACTORS – WITHOUT MATERIALS**
   Enquiries to Labour Only Sub Contractors
   General Procedure for Placing Sub Contract Orders
   BEC Form of Contract

10. **SCAFFOLDING SUB CONTRACTS**
    Enquiries and General Procedure for Placing Orders
    Model Conditions for Hire, Erection and Dismantling of Scaffolding
    (BEC/NASS – 1980 Edition)

11. **TOWER CRANE HIRE CONTRACTS**
    Enquiries and General Procedure for Placing Orders
    Model Conditions for the Hiring of Plant (CPA Conditions – September 1979)

12. **DESIGN WARRANTIES   :   PERFORMANCE BONDS**

    PARENT COMPANY GUARANTEES

    PAYMENTS TO SUB CONTRACTOR'S FOR MATERIALS OFF SITE

APPENDIX 1(b)   *Materials purchasing procedures*

SECTION 2

MATERIALS PURCHASING

Contents:

A   **ORDERING PROCEDURES**

1) General Procedure
2) Requisition Forms
3) Telephone Requisitions
4) Action by Purchasing Office
5) Bulk Call Off Orders
6) Amendment Orders
7) Invoice Queries
8) Nominated Suppliers - JCT 1980 Standard Form
9) Nominated Suppliers - GC/Works/1 Form
10) Named Suppliers
11) Design Warranty - Suppliers
12) Joinery Orders

B   **SCHEDULE OF FORMS**

Schedule of Material Requisition and Supply Forms in Current Use

Terms & Conditions - Bulk Call Off Orders

Additional Terms & Conditions - Ready Mixed Concrete

APPENDIX 1(c)   *Schedule of purchasing forms*

SCHEDULE OF PURCHASING FORMS (PF REFERENCE) IN USE

PF1	Material Supply Enquiry Form
PF2	Material Requisition
PF3	Supply Order
PF3S	Supply Order (site) with £50.00 limit
PF3A	Supply Order Amendment Form
PF22	Sub Contractor's Order Form
PF58	Instruction Order Form
PF100	Sub Contractor's Conditions of Order
PF101	Labour Only Sub Contractor's Conditions of Order
PF102	Sub Contractor's Enquiry Form
PF103	Scaffolding Sub Contractor's Conditions of Order
PF104	Sub Contractor's Acknowledgement of Order
PF105) PF106)	Not Allocated
PF107	Tower Crane Enquiry Form
PF108	Standard Conditions for Hire of a Tower Crane
PF109	Acknowledgement of Tower Crane Order
PF110	Not Allocated
PF111	Not Allocated
PF112	Not Allocated
PF113	Not Allocated
PF114	Scaffolding Enquiry Form
PF115	Standard Specification for Scaffolding
PF116	Acknowledgement of Scaffolding Order
PF117	Specification and Schedule of Scaffolding
PF117A	Specification and Schedule of Scaffolding continuation sheets
PF118	Not Allocated
PF119	Not Allocated
PF120	Supplier's Design Warranty (under hand)
PF121	Supplier's Design Warranty (under seal)
PF122	Sub Contractor's Design Warranty (under hand)
PF123	Sub Contractor's Design Warranty (under seal)
PF124	Sub Contractor's Performance Bond
PF125	Parent Company Guarantee from Sub Contractor
PF126	Supplemental Agreement for Payment of Materials Off Site

Section 2
DELAYS, DISRUPTION AND REIMBURSEMENT

This section is concerned with:
- damages for non-completion of the works by the contractor
- extension of time to which the contractor may be entitled because the works are delayed due to specified relevant events
- matters affecting the regular progress of the works
- direct loss and/or expense caused by matters affecting the regular progress of the works
- obtaining reimbursement.

JCT 80 is used as the 'model' form of contract to demonstrate application of the principles but the various forms of contract contain different conditions with regard to the matters considered in this section. The practitioner must, therefore, ensure that the conditions relevant to the contract in hand are understood and that the firm's procedures are adapted accordingly.

2.1 Site possession by contractor, completion and postponement

The broad picture is that the contractor takes possession of the site on a given date and he undertakes to complete the works on or before the 'Completion Date' (JCT 80, cl 23.1). The actual completion date is agreed when the employer and contractor enter into their contract and the agreed date is written into the appendix to JCT 80.

If the contractor fails to complete the works by the completion date the architect issues a certificate to that effect (JCT 80, cl 24.1).

Issue of the certificate referred to above makes the contractor liable to pay the employer liquidated damages.

The architect may, however, issue instructions postponing any of the work.

Liquidated damages are intended as a means of compensating the employer for any loss which he may incur as a result of the contractor failing to complete on or before the completion date. In the appendix to JCT 80 there is an item 'liquidated and ascertained damages ... at the rate of £ ...... per ......'. When the employer and contractor enter into their contract they fill in the blanks. The sum inserted must be reasonable. It must represent the damages which the employer will actually suffer by way of loss of rental income, increased costs of financing the project, etc, as a result of the contractor's failure. English courts have never held that liquidated damages should be treated as a means of penalising the contractor. The blank spaces in the appendix are usually completed as £x per day, week or month as the parties consider appropriate.

Liquidated damages often involve large sums so contractors are understandably concerned that the architect does not issue a certificate to the effect that he, the contractor, has failed to complete the works by the completion date.

Relevant events There is provision in JCT 80 for the architect to make an extension of time if the progress of the works is delayed as a result of a number of 'Relevant Events' (JCT 80, cl 25.1). Twelve relevant events are listed in JCT 80, clause 25.4. They include exceptionally adverse weather conditions, the contractor not having received in due time necessary instructions and delay on the part of nominated sub-contractors or nominated suppliers. NSC/4, the nominated sub-contractor form, has a similar list of relevant events in clause 11.

Delay Clause 25.2.1.1 of JCT 80 requires that if it becomes reasonably apparent that the progress of the works is being or is likely to be delayed the contractor shall give written notice to the architect of the 'material circumstances including the cause or causes of the delay and identify in such notice any event which in his opinion is a relevant event'. Where the material circumstances in the written notice include reference to a nominated sub-contractor the contractor is required to send a copy of the notice to the nominated sub-contractor (JCT 80, cl 25.2.1.2).

The contractor's written notice should include particulars of the expected effects of the relevant event to which he refers in his notice and he should also estimate the extent of the expected delay in the completion of the works beyond the completion date (JCT 80, cl 25.2.2.2). If it is not practicable for the contractor to give the above particulars at the time he gives his written notice he is required to do so as soon as possible after the notice and he is also required to send copies of the particulars to nominated sub-contractors.

Particulars kept up to date Clause 25.2.2.3 provides for the particulars and estimate to be kept up-to-date and the contractor is required to give to the architect further written notices as may reasonably be necessary or as the architect may reasonably require. Here, again, copies must be sent to nominated sub-contractors.

Written extension of time Having received from the contractor any notices and particulars and estimates the architect is obliged to decide if, in his opinion, any of the events stated by the contractor to be the cause of the delay is a relevant event and if the completion of the works is likely to be delayed by it. If so, the architect is obliged to give to the contractor, in writing, an extension of time by fixing such later date as the completion date as he, the architect, estimates to be fair and reasonable (JCT 80, cl 25.3.1). In his letter to the contractor the architect is obliged to state which of the relevant events he has taken into account and the extent to he has had regard to any instruction which he may have given requiring as a variation the omission of any work issued since the fixing of the previous completion date.

Action by architect within 12 weeks The architect is required to take the action described above within 12 weeks from receipt of the notice, particulars, etc, from the contractor or, if the completion date is less than 12 weeks away, the architect must act before the completion date (JCT 80, cl 25.3.1).

Two points, in particular, emerge from the foregoing requirements. The first is that the architect has the power to try to reduce the effect of delays which have arisen by omitting part of the works. The second is that he is required to take action (to fix a new completion date) within a stated period of time.

The action expected of the architect in clause 25.3.1 is, incidentally, referred to as the 'first exercise of his duty' so presumably the following action, set out in clause 25.3.2, may be regarded as his second exercise.

Sub-contractors' obligations Reference is made in section 1.6 to the sub-contractor's obligation to carry out and complete the sub-contract works and extension of time are set out in clause 11 of NSC/4.

The (nominated) sub-contractor undertakes to:

'carry out and complete the Sub-Contract Works in accordance with the agreed programme details in the Tender, Schedule 2, item 1C, and reasonably in accordance with the progress of the Main Contract Works but subject to receipt of the notice to commence work on site as detailed in the Tender, Schedule 2, item 1C, and to the operation of Clause 11.2'. (NSC/4, cl 11.1)

Clause 11.2 has the heading 'Extension of Sub-Contract time'.

Sub-contractor's written notice NSC/4, clause 11.2 describes the action which should be taken if the sub-contract works are delayed. The clause reads:

11.2.1.1 If and whenever it becomes reasonably apparent that the commencement, progress or completion of the Sub-Contract Works or any part thereof is being or is likely to be delayed, the Sub-Contractor shall forthwith give written notice to the Contractor of the material circumstances including the cause or causes of the delay and identify in such notice any matter which in his opinion comes within clause 11.2.2.1. The Contractor shall forthwith inform the Architect of any written notice by the Sub-Contractor and submit to the Architect any written representations made to him by the Sub-Contractor as to such cause as aforesaid.

Earlier completion date This clause enables the architect to fix an earlier completion date than that previously fixed if in his opinion the fixing of such earlier completion date is fair and reasonable having regard to any instructions he may have given requiring as a variation the omission of any work. The instruction which gives rise to the omission must have been given after the last occasion on which the architect made an extension of time and (cl 25.3.6) no decision by the architect under clause 25.3 shall fix a completion date earlier than the date for completion stated in the appendix to JCT 80.

Not later than the expiry of 12 weeks from the date of Practical Completion the architect has three courses of action which he may take as a sweeping-up exercise. These are:

– he may fix a completion date later than that previously fixed if in his opinion this would be fair and reasonable (JCT 80, cl 25.3.3.1)

– he may fix a completion date earlier than that previously fixed if in his opinion ... etc (JCT 80, cl 25.3.3.2)
– he may confirm to the contractor the completion date previously fixed (JCT 80, cl 25.3.3.3).

He must not just let matters rest – he must take one of the courses.

It should be noted that 'the Architect shall notify in writing to every Nominated Sub-Contractor each decision of the Architect under clause 25.3 fixing a Completion Date'.

All three parties are involved; the sub-contractor is required to give written notice to the contractor who is obliged to inform the architect and pass on the sub-contractor's 'representations'.

If it is practicable for the sub-contractor to provide all the 'particulars' which the contractor and architect need to know in order to exercise their respective duties he should provide these particulars with his written notice. If he is not able to do so at that time he should do so 'as soon as possible after such notice'.

Clause 11.2 requires the sub-contractor to submit his particulars and estimate to the contractor.

The wording of the above sub-clauses is very similar to that contained in JCT 80 when describing the action which the contractor should take. The contractor is obliged to submit to the architect the particulars and estimate referred to above and 'if so requested by the Sub-Contractor, join with the Sub-Contractor in requesting the consent of the Architect under Clause 35.14 of the Main Contract Conditions'. This takes us back to JCT 80 where we find:

35.14.1 The Contractor shall not grant to any Nominated Sub-Contractor any extension of the period or periods within which the Sub-Contract Works (or where the Sub-Contract Works are to be completed in parts any part thereof) are to be completed except in accordance with the relevant provisions of Sub-Contract NSC/4 or NSC/4a as applicable which requires the written consent of the Architect to any such grant.

35.14.2 The Architect shall operate the relevant provisions of Sub-Contract NSC/4 or NSC/4a as applicable upon receiving any notice particulars and estimate and a request from the Contractor and any Nominated Sub-Contractor for his written consent to an extension of the period or periods for the completion of the Sub-Contract Works or any part thereof as referred to in clause 11.2.2 of Sub-Contract NSC/4 or NSC/4a as applicable.

The main contract and the sub-contract conditions are mutually compatible.

JCT 80 requires the contractor not to grant any extension of time to the sub-contractor without the written consent of the architect (JCT 80, cl 35.14.1) but the architect is obliged to operate the relevant provision of NSC/4.

NSC/4 states the action which the architect must take.

We are, again, dependent upon the architect's 'opinion' regarding the extent to which the sub-contractor has caused delay and any extension of time which is to be given.

The action to be taken by the parties is stated in NSC/4 clause 11.2.

11.2.2 If on receipt of any notice, particulars and estimate under clause 11.2.1 and of a request by the Contractor and the Sub-Contractor for his consent under clause 35.14 of the Main Contract Conditions the Architect is of the opinion that:
.2.1 any of the matters which are stated by the Sub-Contractor to be the cause of the delay is an act, omission or default of the Contractor, his servants or agents or his sub-contractors, their servants or agents (other than the Sub-Contractor, his servants or agents) or the occurrence of a Relevant Event; and
.2.2 the completion of the Sub-Contract Works or any part thereof is likely to be delayed thereby beyond the period or periods stated in the Tender, Schedule 2, item 1C, or any such revised period or periods, then the Contractor shall, with the written consent of the Architect, give an extension of time by fixing such revised or further revised period or periods for the completion of the Sub-Contract Works or any part thereof as the Architect in his written consent then estimates to be fair and reasonable.

It will be noted that, in addition to the 'occurrence of a Relevant Event', clause 11.2.2.1 refers to 'an act, omission or default of the Contractor...' as a possible cause of delay.

In clause 11.2.2 we have the rather quaint, three-sided arrangement where it is the architect who considers the sub-contractor's notice, particulars and estimate which he, the architect, received via the contractor and which he (the architect again) must consent to (if he is of the opinion that he should do so) via the contractor.

The remainder of the clause reads:

.2.3 which of the matters including any of the Relevant Events, referred to in clause 11.2.2.1 they have taken into account; and
.2.4 the extent, if any, to which the Architect, in giving his written consent, has had regard to any instruction requiring as a Variation the omission of any work issued under clause 13.2 of the Main Contract Conditions since the previous fixing of any such revised period or periods for the completion of the Sub-Contract Works or any part thereof,

and shall, if reasonably practicable having regard to the sufficiency of the aforesaid notice, particulars and estimate, fix such revised period or periods not later than 12 weeks from the receipt by the Contractor of the notice and of reasonably sufficient particulars and estimates, or, where the time between receipt thereof and the expiry of the period or periods for the completion of the Sub-Contract Works or any part thereof is less than 12 weeks, not later than the expiry of the aforesaid period or periods.

That completes the first exercise by the contractor of the duty which, as may be seen above, is similar to that performed by the architect in respect of the contractor's notices. We have now reached the point where the contractor, in agreement with the architect, has fixed revised periods and stated which matters have been taken into account and the extent to which the architect has had regard to any instructions he may have given requiring as a variation the omission of any work.

The contractor, it will be noted, is required to fix the revised period or periods not later than 12 weeks from his receipt of the notice, etc, from the sub-contractor or not later than the expiry of the period/s for the completion of the sub-contract works. This requirement may place the contractor in some difficulties because if experience from the past is to be used as a guide it is likely that both contractor and sub-contractor will be giving notices of delay at much the same times and probably in respect of a number of different relevant events. It may be difficult for the contractor to fix a revised period not later than 12 weeks from receipt of the sub-contractor's notice if that notice (together with particulars and estimate) has to be considered by the architect who, in turn, has 12 weeks to 'exercise his duty'.

Shortened period for completion JCT 80 appears to take into account the possibility of 'time' being important to the employer. There is, therefore, provision in clause 11.2.3 for the architect to issue an instruction which requires as a variation the omission of work. Presumably the thinking is that if there is less work to be done it will take less time and the period for completion of the sub-contract works may therefore be shortened. The arrangement in this respect is similar in both JCT 80 and NSC/4.

There is an opportunity for 'second thoughts' on the part of the contractor and architect with regard to lengthening, shortening or confirming the period or periods for completion of the sub-contract works not later than the expiry of 12 weeks from the date of practical completion of the sub-contract works or from the date of practical completion of the main contract, whichever occurs first.

Relevant events It was stated above that the contractor and/or the sub-contractor require an extension of time because without it they may be liable to pay the employer liquidated damages. There is, however, a clearly defined list of events, named 'Relevant Events' for which an extension of time may be given. The relevant events are set out in clause 25 of JCT 80 and in clause 11.2.5 of NSC/4. The lists are very similar.

JCT 80 clause 25.4 contains the following items. The clause begins:

25.4 The following are the Relevant Events referred to in clause 25 and the first relevant event is:
25.4 .1 force majeure.

'Force majeure' is, literally, 'forces beyond one's control', and is a term used in legal documents to embrace a wide range of major events some of which are included as other relevant events in clause 11.2.5. Generally, however, acts of God, war, earthquakes and the like would come in the force majeure category.

The relevant event referred to in clause 25.4.2 reads:

25.4 .2 exceptionally adverse weather conditions.

The key word is 'exceptionally'. Contractors and sub-contractors alike must accept the fact that adverse weather conditions are, in Britain, the rule rather than the exception and that to qualify as a relevant event the weather conditions must be *exceptionally* adverse. It is, in practice, difficult for a contractor to establish when the adversity of the weather conditions becomes exceptional because so many factors are involved. It may be, for instance, that adverse weather conditions, however exceptionally adverse, would have little or no effect on a project which had advanced to a point where the building was weathertight. On the other hand weather conditions would not have to be even exceptionally adverse before a foundation engineering or piling sub-contractor working on a heavy clay soil found it impossible to carry out his work.

How then may one measure the point

beyond which the weather conditions become exceptionally adverse?

One source of information on which measurements may be based is the Meteorological Office records. By comparison of records over a period of time it may be possible for the contractor or sub-contractor to determine exceptions from the norm which will satisfy the architect that the adverse weather conditions which have been experienced on the project under consideration have been exceptional and thus qualify as a relevant event. These records are an imprecise method of measurement because they do not normally take into account the time of the day or night when the conditions occurred nor do they take into account the nature of the site but they are, at least, an indication of condition and they may provide a starting point from which a more detailed 'case' may be prepared.

The relevant event referred to in clause 25.4.3 reads:

25.4 .3 loss or damage occasioned by any one or more of the 'Specified Perils'.

'Specified Perils' are those against which the contractor must insure to comply with clause 22 of JCT 80. They appear in Part 1 of JCT 80 in the definitions and they make a formidable list which starts with fire, lightning, explosion, storm and includes overflowing water tanks, articles dropped from aircraft and civil commotion.

The relevant event referred to in clause 25.4.4 reads:

25.4 .4 civil commotion, local combination of workmen, strike or lock-out affecting any of the trades employed upon the Works or any of the trades engaged in the preparation, manufacture or transportation of any of the goods or materials required for the Works.

It is important for the contractor to note that the clause refers not only to men employed upon the works but also to trades engaged in the preparation, manufacture or transportation of any of the goods or materials required for the works. It will almost certainly be necessary for the contractor to obtain appropriate information relating to off-site strikes in order to demonstrate to the architect and contractor the effect of such strikes on the project as a whole.

The relevant event referred to in clause 25.4.5 reads:

25.4 .5 compliance with the Architect's instructions;
    .5 .1 under clauses 2.3, 13.2, 13.3, 23.2, 34, 35 or 36; or
    .5 .2 in regard to the opening up for inspection of any work covered up or the testing of any of the work, materials or goods in accordance with clause 8.2 (including making good in consequence of such opening up or testing) unless the inspection or test showed that the work, materials or goods were not in accordance with this Contract.

The clauses referred to in clause 25.4.5.1 are concerned with:

2.3   discrepancies in or divergences between documents
13.2 instructions requiring a variation
13.3 instructions on provisional sums
23.2 instructions regarding postponement
34   the effect of finding antiquities
35 and 36 nominated sub-contractors and nominated suppliers

The relevant event in clause 25.4.6 reads:

25.4 .6 the Contractor not having received in due time necessary instructions, drawings, details or levels from the Architect for which he specifically applied in writing provided that such application was made on a date which having regard to the Completion Date was neither unreasonably distant from nor unreasonably close to the date on which it was necessary for him to receive the same.

A most important aspect of this relevant event, frequently overlooked by contractors (and sub-contractors), is that they cannot just sit back and blame the architect and expect him to give them extensions of time because they have not received in due time necessary instructions, etc, *unless* (and it is an important 'unless') they have specifically applied in writing for the instructions, etc, on a date which having regard to the completion date, etc, was neither unreasonably distant from nor unreasonably close to the date on which it was necessary for them, the contractors and sub-contractors, to receive them. At what point is a date unreasonably distant from or close to?

Provided the contractor is able to explain, if asked by the architect, why he needs the instructions, etc, by a certain date, it is most unlikely that his application will be regarded as unreasonable. It is the contractor who has to carry out and complete the contract works and only he is able to say when it will be necessary for him to place orders for this or that piece of equipment or commence off-site manufacture of part of the contract work.

The intention of restricting the timing of the applications for instructions, etc, appears to be to assist the architect or design team to plan their design work and order their priorities.

There has been considerable discussion about whether a contractor (or sub-contractor) would be considered to meet the timing requirement if, when he entered into contract, he were immediately to provide a schedule of dates, covering a period some months ahead showing when he required instructions, etc. for various aspects of his work. For practical purposes the instruction dates may be more conveniently presented in the form of a bar-chart which relates the dates by which instructions should be given to the dates when the work is to be carried out. Is his application, which is being made far in advance of the time when he actually requires the drawings, etc, 'unreasonably distant from'?

It will be noted from the wording of the clause that it is the date of the application which has the time restriction.

When considering points of contract and procedures there is something to be said for putting oneself in the position of an arbitrator who has been called upon to make an award. Would he have to take into account the contractual as well as the practical aspects? The answer is, of course, that he most certainly would. The contractor would, then, be well advised to be on the safe side. He will make his own plans when he enters his contract so he should acquaint the architect with his requirements at that time in the form of an application. The contractor's schedule may then provide the basis for subsequent 'applications' as the project progresses. If the works are delayed, the contractor should advise the architect that the actual dates by which instructions should be given may be varied from those shown on the schedule, provided that the times between instructions and commencements on site are maintained as shown on the schedule. Alternatively, the contractor may have committed a sub-contractor to keep a certain period of time free in his workshop, in which event it would be reasonable for the contractor to insist that the architect adheres to the date shown on the schedule.

Clause 25.4.6 is important to the contractor for two reasons. The first is that the relevant event to which it refers may provide him with an extension of time, and the second that such an extension of time may lead to reimbursement by reason of the regular progress of the contract works being materially affected.

The matter of reimbursement is discussed below.

The relevent event in clause 25.4.7 reads:

25.4 .7 delay on the part of Nominated Sub-Contractors or Nominated Suppliers which the contractor has taken all practicable steps to avoid or reduce.

It is important for the contractor to ensure that he takes 'all practicable steps to avoid or reduce the delay'. An extension of time does not automatically follow the delay caused by the sub-contractor or supplier.

The relevant event in clause 25.4.8 reads:

25.4 .8 .1 the execution of work not forming part of this Contract by the

Employer himself or by persons employed or otherwise engaged by the Employer as referred to in clause 29 or the failure to execute such work;

.8 .2 the supply by the Employer of materials and goods which the Employer has agreed to provide for the Works or the failure so to supply.

The contractor has little or no control over the 'persons' or supply by the employer of materials and goods. It may be noted that the contractor is not required to have taken 'all practicable steps to avoid or reduce the delay caused'. This distinguishes between the nominated sub-contractors and suppliers and the persons referred to in clause 29.

The relevant event in clause 25.4.9 reads:

25.4 .9 the exercise after the Base Date by the United Kingdom Government of any statutory power which directly affects the execution of the Works by restricting the availability or use of labour which is essential to the proper carrying out of the Works or preventing the Contractor from, or delaying the Contractor in, securing such goods or materials or such fuel or energy as are essential to the proper carrying out of the Works.

This clause provides the contractor with protection against the effect of government actions. The 'Base Date' is the date stated in the appendix to the conditions of contract. The contractor's tender was based on prices prevailing on that date.

The relevant event in clause 25.4.10 reads:

25.4. .10 .1 the Contractor's inability for reasons beyond his control and which he could not reasonably have foreseen at the Base Date to secure such labour as is essential to the proper carrying out of the Works; or

.10 .2 the Contractor's inability for reasons beyond his control and which he could not reasonably have foreseen at the Base Date to secure such goods or materials as are essential to the proper carrying out of the Works.

Perhaps 'reasonably have foreseen' are the key words in this clause. It is not sufficient for the contractor to claim that labour or goods are unavailable. Conditions must have changed since he submitted his tender.

The relevant event in clause 25.4.11 reads:

25.4 .11 the carrying out by a local authority or statutory undertaker of work in pursuance of its statutory obligations in relation to the Works, or the failure to carry out such work.

Statutory undertakers are the various 'boards', which were formed to provide electricity, gas, water, etc.

The contractor (and sub-contractor) is entitled to an extension of time in respect of the clause only if the delay is caused by the local authority or statutory undertaker when executing its statutory obligations. If they are acting as nominated sub-contractors, as would be electricity boards which undertake electrical installation works, an application for an extension of time by a contractor should be under the clause 25.4.7 – the relevant event which refers to nominated sub-contractors.

The relevant event in clause 25.4.12 reads:

25.4 .12 failure of the Employer to give in due time ingress to or egress from the site of the Works or any part thereof through or over any land, buildings, way or passage adjoining or connected with the site and in the possession and control of the Employer, in accordance with the Contract Bills and/or the Contract Drawings, after receipt by the Architect of such notice, if any, as the Contractor is required to give, or failure of the Employer to give such ingress or egress as otherwise agreed between the Architect and the Contractor.

The failure to give due 'ingress to or egress from' must refer to matters not included in the contract documents. It may be, for example, that the contract documents require work to be carried out in a prescribed sequence which determined ingress and egress. The employer's failure must involve a change from the conditions stated in the contract.

The last relevant event reads:

25.4 .13 where clause 23.1.2 is stated in the Appendix to apply, the deferment by the Employer of giving possession of the site under clause 23.1.2.

Clause 23.1.2 states that where the clause is stated in the appendix to apply, the employer may defer giving possession for a period not exceeding six weeks or a lesser period stated in the appendix.

The relevant events are the events which entitle the contractor to an extension of time and thus relieve him from liability to pay liquidated damages. In addition to the relevant events, the discovery of antiquities may provide such an entitlement.

Antiquities Recognition by JCT of the effect of discovery of, in particular, antiquities on construction sites in terms of disruption of the regular progress of the works and consequential direct loss and expense to the contractor led to the present clause 34.

If the contractor discovers on site 'fossils, antiquities and other objects of interest or value', he is obliged to endeavour not to disturb the object and to cease work which would endanger the object or prevent or impede its excavation or its removal. He must, also, take steps to preserve the object in the position in which it was found and inform the architect or clerk of works (cl 34.1).

The initiative then rests with the architect to issue instructions in regard to what is to be done concerning the object. He may instruct the contractor to permit the examination, excavation or removal of an object 'by a third party' – typically an archaeologist. The third party is deemed to be a 'person' for whom the employer is responsible to which reference is made above (cl 34.2).

If in the opinion of the architect compliance with clause 34.1 or 34.2 involves the contractor in direct loss and/or expense the architect is obliged to ascertain or instruct the quantity surveyor to ascertain the amount of the loss and/or expense (cl 34.3).

The architect is also obliged to state in writing to the contractor what extension of time, if any, he has made. Further reference to this is made in section 2.

The contractor should implement procedures to record any disturbance to the regular progress of his work caused by discovery of antiquities or arising from subsequent architect's instructions so that he may recover any loss and/or expenses he may incur.

It is important for the contractor to remember that being given an extension of time does not entitle him to reimbursement for any direct loss and/or expense caused to him. Direct loss and/or expense are matters for clauses 8.3, 13.5.5, 26 and 34.

Clause 26.1 sets out the procedures to be followed by contractor, architect and quantity surveyor should the regular progress of the works be materially affected.

26.1 If the Contractor makes written application to the Architect stating that he has incurred or is likely to incur direct loss and/or expense in the execution of this Contract for which he would not be reimbursed by a payment under any other provision in this contract due to deferment of giving possession of the site under clause 23.1.2 where clause 23.1.2 is stated in the Appendix to be applicable or because the regular progress of the Works or of any part thereof has been or is likely to be materially affected by any one or more of the matters referred to in clause 26.2; and if as soon as the Architect is of the opinion that the direct loss and/or expense has been incurred or is likely to be incurred due to any such deferment of giving possession or that the regular progress of the works or of any part thereof has been or is likely to be so materially affected as set out in the application of the Contractor then the Architect from time to time thereafter shall ascertain, or shall instruct the Quantity Surveyor to ascertain, the amount of such loss and/or expense which has been or is being incurred by the Contractor; provided always that:

26.1 .1 the Contractor's application shall be made as soon as it has become, or should reasonably have become apparent to him that the regular progress of the Works or of any part thereof has been or was likely to be affected as aforesaid, and

26.1 .2 the Contractor shall in support of his application submit to the Architect upon request such information as should reasonably enable the Architect to form an opinion as aforesaid, and

26.1 .3 the Contractor shall submit to the Architect or to the Quantity Surveyor upon request such details of such loss and/or expense as are reasonably necessary for such ascertainment as aforesaid.

2.2 Loss and expense

The margin note opposite JCT 80, clause 26.1 reads:

'matters materially affecting regular progress of the Works – direct loss and/or expense.

The note includes all the key words.

There is a 'list of matters' in clause 26.2 for which, if they materially affect the regular progress of the Works, the contractor is entitled to seek reimbursement for direct loss and/or expense which he has suffered.

The 'written application' is what is frequently described as a 'claim' but that word is not used in JCT 80, clause 26 nor in NSC/4, clause 13, the sub-contract counterpart, in this context.

As far as JCT 80, clause 26.1 is concerned the position is simply that the contractor makes a written application stating that 'he has incurred or is likely to incur direct loss and/or expense...(etc)' and provided the architect 'is of the opinion that the regular progress of the works...is likely to be so materially affected as set out in the application' he, the architect, shall ascertain the amount of such loss and/or expense.

Alternatively the architect may instruct the quantity surveyor to ascertain the amount.

Clause 26.1 requires that the contractor's application must be timely, that he must submit any information and details of such loss and/or expense as the architect or quantity surveyor requests. These provisions are essentially the same as those which apply to the sub-contractor.

List of matters The matters affecting the regular progress of the Works for which the contractor may be reimbursed are listed in clause 26.2. The list reads:

26.2 .1 the Contractor not having received in due time necessary instructions, drawings, details or levels from the Architect for which he specifically applied in writing provided that such application was made on a date which having regard to the Completion Date was neither unreasonably distant from nor unreasonably close to the date on which it was necessary for him to receive the same;

26.2 .2 the opening up for inspection of any work covered up or the testing of any of the work, materials or goods in accordance with clause 8.3 (including making good in consequence of such opening up or testing), unless the inspection or test showed that the work, materials or goods were not in accordance with this Contract;

26.2 .3 any discrepancy in or divergence between the Contract Drawings and/ or the Contract Bills and/or the Numbered Documents;

26.2 .4 .1 the execution of work not forming part of this Contract by the Employer himself or by persons employed or otherwise engaged by the Employer as referred to in clause 29 or the failure to execute such work;

26.2 .4 .2 the supply by the Employer of materials and goods which the Employer has agreed to provide for the Works or the failure so to supply;

26.2 .5 Architect's instructions under clause 23.2 issued in regard to the postponement of any work to be executed under the provisions of this Contract;

26.2 .6 failure of the Employer to give in due time ingress to or egress from the site of the Works, of any part thereof through or over any land, buildings, way or passage adjoining or connected with the site and in the possession and control of the Employer, in accordance with the Contract Bills and/or the Contract Drawings, after receipt by the Architect of such notice, if any, as the Contractor is required to give, or failure of the Employer to give such ingress or egress as otherwise agreed between the Architect and the Contractor;

26.2 .7 Architect's instructions issued under clause 13.2 requiring a Variation or under clause 13.3 in regard to the expenditure of provisional sums (other than work to which clause 13.4.2 refers).

A comparison of the list of relevant events with the list of matters (for no apparent reason the term 'Relevant Events' warrants initial capital letters but 'matters' does not) reveals that there are 13 relevant events and seven matters. Clearly, not all 'relevant events' are 'matters' leading to an entitlement to recover loss and/or expense.

When delays occur and the regular progress of the works has been materially affected there are frequently several concurrent causes. It is important for a contractor to obtain an extension of time but whilst this may relieve him of the possibility of a claim against him by the employer for liquidated damages it will not reimburse him for loss and/or expense. If, then, the contractor is faced with giving written notice of more than one relevant event which is causing delay he may consider it prudent to concentrate on the relevant event/s which is/are included in the list of matters in clause 26.2 rather than the event/s which is/are not. There is, however, a risk which the contractor should have in mind and this is that if he is not given an extension of time in respect of the item/s of which he has given notice he may find it difficult to introduce a new item at a later date.

It is important for the contractor to remember that he must give a written *notice of delay to progress* (for which he seeks an extension of time and, thus, relief from a claim for liquidated damages), and he must make a written *application that he has incurred loss and/or expense*. The notice and the application are separate, discrete tasks. This is not to say that the tasks may not be combined in a single letter but, if they are, it must be made clear that both notice and application are involved.

2.3 Claims for direct loss and/or expense

A claim is a 'demand for something as due... right, title...' (OED)

The right arises from the contract entered into by the parties. Should they fail to reach agreement, so that it is necessary for one to assert his right, he must do so in law. Most building contracts provide for disputes to be settled by arbitration, so a claim should be prepared with a view to it being used by an arbitrator. Indeed, as an arbitrator may call for almost any documents, calculations, correspondence, etc, etc, to be 'discovered' (disclosed) by the parties (see section 4.6), the contractor should take the greatest care with all his documentation avoiding inaccuracies, ambiguities and (in correspondence) emotional statements which might detract from his equanimity in the eyes of a reader, such as the arbitrator, at a later date when day-to-day pressures are forgotten.

It should not be necessary for a contractor to make a claim. Where certain matters are affecting the regular progress of the works he may make a 'written application' that he is 'likely to incur direct loss and/or expense', etc. It is then up to the architect to ascertain the loss, etc, or to instruct the quantity surveyor to do so. Where the work has been varied the quantity surveyor is responsible for valuing the variations.

What reason, then, has the contractor to prepare a claim?

It must be remembered that the contractor is obliged to submit details of his loss and/or expense to the architect or quantity surveyor if requested to do so and most contractors are aware that applications for reimbursement are not always recognised by architects, so that the contractors find it necessary to demonstrate the extent of their loss. As far as variations are concerned, the contractor has a right to refer any dispute to arbitration so the sub-contractor, too, must be able to assess the value of his work for himself.

Having established a need for the contractor to be able to demonstrate his loss and make valuations we may consider ways and means.

Preparing a claim A prudent contractor prepares the ground for a future claim when he compiles his estimate. He does this by recording the basis of his estimate, his calculations, etc, so that he is able to show the extent of the information which was available at the time of tender. He should be able to demonstrate the methods and plans which he intended to use, should his tender be successful. An estimate is only as good as the information available for its preparation. It may be said that the sum of a claim is the difference between cost of the works as known to the contractor at time of tender and the cost of the works in the light of all the information.

Records are important to the contractor throughout the life of all his projects.

The irony is that if his records are good he is less likely to find it necessary to submit a claim. Claims and disputes are usually in inverse ratio to the quality of the sub-contractor's records.

The steps in the preparation of a claim are essentially the same as those used for method or work study or, for that matter, any systematic form of examination, analysis, diagnosis, etc.

Selection of matters The range of matters which may provide the subject of a claim will be largely determined by the 'list of matters' in JCT 80 clause 26 or by the contractor's default or omission. The contractor should not be too selective when considering which items to include because there will almost certainly be a degree of negotiation with the architect or quantity surveyor in due course and the contractor may feel it prudent to have some 'face-savers' at that time.

Nonetheless, the contractor should not work on the 'settle-for-half' principle – merely include some spare capacity. Experience with negotiations of this sort often reveals that the items which were included almost as make-weights are those which are most readily accepted by the architect.

The selection of items should emerge from the initial reading of the files which the person preparing the claim must undertake. There is frequently no clear line between individual items.

Sources of information include: estimate build-up, tender data, contract documents, specifications, bills of quantities, contract drawings, drawing schedules, working drawings, drawings with free-hand sketches by the architect or consultants used to illustrate instructions, architect's instructions and variation orders, contractor's directions, correspondence with all parties to the contract, minutes of meetings, records of telephone and other conversations, foreman's and site manager's reports or diaries, programmes, manpower planning histograms, estimates from and orders placed with suppliers and sub-contractors, delivery vouchers, plant and tool schedules, labour and plant daily allocation sheets, Meteorological Office records, progress record photographs, 'special event' photographs, applications for payment, quantity surveyor's recommendations and architect's certificates, codes of practice, British Standard Specifications, reports by Royal Commissions, etc, etc.

The above are all prospective sources of information for claims. The list is only indicative of the wealth of information which may prove useful – in effect, anything and everything.

Recording the facts This is usually a matter of scrap-booking cuttings from the contractor's files. Original documents should be photocopied. A single document may contain material for several separate items so a system of coded slips may provide a method of arranging events by 'item' in chronological order.

Size A4 sheets guillotined into four narrow slips with holes punched in the top left-hand corners for treasury tags make it possible for events to be shuffled into order and for additional events to be inserted. The copies of a letter or site meeting minutes, as the case may be, can be pasted on the slips and referenced. If claims for several different projects are being prepared simultaneously the slips may be made from different-coloured papers; yellow for contract A, blue for contract B, etc, or each slip may be marked A, B, etc, near the item no. to prevent confusion.

This may appear an unnecessarily elaborate precaution but a claim may be in course of preparation for several months and a number of people may be involved. The contractor's credibility is undermined if he claims for a contract A event on contract B! If the cuttings are too large to fit on single slips they may be accommodated on continuation slips (if cuttings are three or four paragraphs long) or folded over on a single slip.

As the claim preparation develops it may be expedient to combine one or two items. It is a simple matter to combine slips but not so simple to split them. The object of recording the facts is to facilitate examination and the 'stories' which the records tell are more credible if they are derived from a variety of sources. They should be related to time-scales and bar charts may be useful for this purpose.

The line between 'recording' and 'examining' is blurred. Obviously the recorder will examine the facts to some extent whilst making the record but as far as possible assembling the facts should precede the examination.

Examining the facts This frequently produces surprises for the preparer of a claim who has been involved in a project for much of its life. The items which he/she had, as manager, considered to be primary causes of delay and disturbance of the regular progress of the works appear under examination to be less significant on paper. The converse applies.

The examination should compare anticipation/expectation with actual achievement. The contractor, or his estimator on his behalf, prepares an estimate which provides the basis of his tender from the data which is available at that time. If the circumstances, which the contractor had reason to believe would exist, have changed since he submitted his tender he is entitled to consider the effect of the changes.

Specifications, drawings and programmes should be examined for variations. The dates of revision letters or drawings should be recorded and the actual changes (which should be noted in the legend on the drawing if the architect and consultants have done their jobs properly), should be studied. All drawings which have been revised should have been retained by the contractor and must be compared with subsequent revisions and with the notes in the legend. The importance of a systematic approach to estimating, contract administration and procedures for storing data is now apparent.

The findings are best recorded on barcharts because the time when an event occurred is frequently a crucial factor and the effects of changes can more easily be determined if related to a programme. Figure 2.1 shows how the bar-chart may appear.

See above how the contractor's information requirements may be presented to the architect using the contractor's programme as a foundation. Comparison of figures 2.1 and 2.2 makes it possible to ascertain (and later to demonstrate to the architect) the effect of information flow upon the progress of the works.

As the examination proceeds the contractor will almost certainly find that his original assessments of the situation will change and he may find that he simply does not have sufficient evidence about some items to make it possible for him to present a worthwhile case to the architect. It is not enough for him to know he has suffered. It

that the Contractor shall not be entitled so to claim

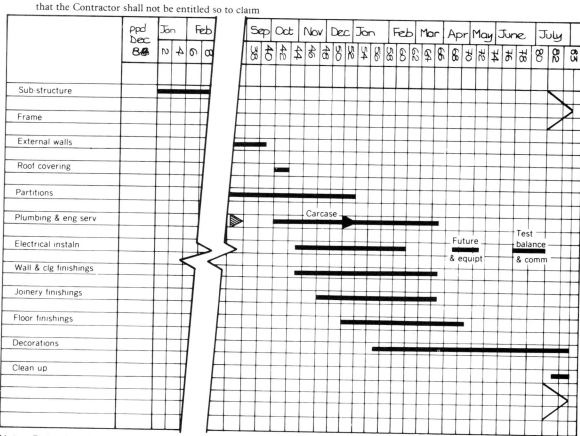

Note: Dates by which all instructions, drawings, etc, regarding electrical works must be issued are shown ▷ for carcase ▶ for fixtures and equipment

2.1 *Application for instructions, drawings, etc*

does not matter how sad the story his cost records may show; if his claim is to succeed he must be able to demonstrate that he has complied with the contract conditions in all the ways discussed. The most experienced claims 'expert' cannot make bricks without straw.

Analysis or diagnosis This is the task which should follow examination but in practice some measure of analysis will almost certainly occur whilst the examination is in progress.

The contractor's analysis of the data which he has recorded, collected and col-lated will lead him to decide on the form which his claim will take or if he has a claim to make!

He analyses facts rather than fantasies and he should attempt to be objective however difficult this may be for someone who frequently has been intimately involved with the project and who may stand to lose (or gain) substantial sums. One way in which a contractor may increase his objectivity is for him to put himself in the place of an arbitrator considering the evidence placed before him with a view to making an award.

If he considers an item to have merit it is

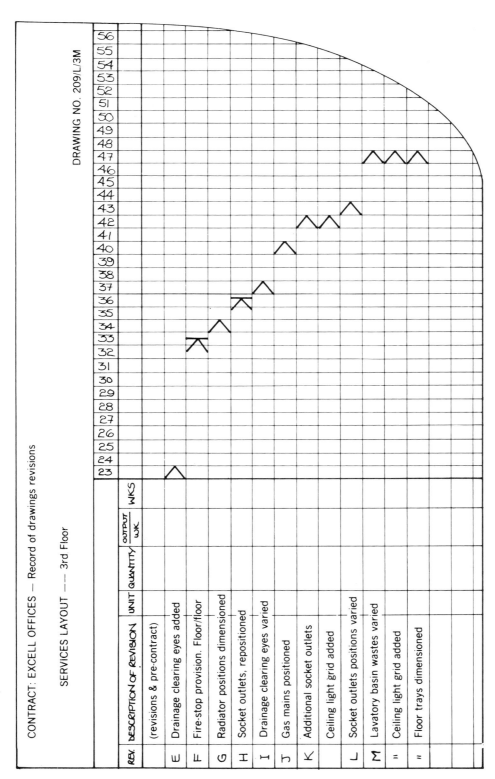

2.2 *Recording revisions of drawings*

worth including it in the claim; if not he should consider dropping it or, perhaps, strengthening it if this is possible. The basis of his analysis must be the contract conditions. The contractor must have done all that he ought to have done if his claim is to succeed.

Format of claim The format for a claim will depend to a considerable extent on its size and complexity but for the majority of contractor's claims the principal headings may be:

(a) contract particulars
(b) statement of events leading to claim
(c) statement of claim
(d) evaluation of claim
(e) appendices.

Contract particulars These should identify:

(a) the title of the project
(b) the parties to main contract and sub-contracts (client, architect, main contractor, quantity surveyor, other consultants, and sub-contractors)
(c) the tender
(d) specification or bills of quantities
(e) drawings, with details of numbers and revision letters
(f) articles of agreement (form of contract with details of amendments, deletions, etc).

Statement of events leading to claim This statement will probably open with a common theme indicating the manner in which the project progressed and the 'communication gap' widened. To facilitate cross-referencing the statement should be itemised. Dates and sources of information should be given.

The 'facts' recorded on the slips which were discussed above are used for this purpose.

Entries in the statement of events might read:

2.18 Minutes of Site Meeting 9, (9.03)
10 Feb. record that the services programme for the kitchen had been received and that a delay of three weeks had already occurred. 9.14 records that joist straps are still needed.

2.19 Letter... enclosed drawing B/72/
10 Feb. 105 detailing holes through roof slab.

When the common theme has been established the contractor is able to record the events related to individual items.

The statement of events should provide irrefutable evidence of the disruption (delay or whatever) which the contractor has experienced so that a succinct 'statement of claim' may be prepared.

The entries should be 'factual' and capable of standing on their own. In some instances, however, they may require amplification by means of interpolated paragraphs.

Such paragraphs must be equally factual but they are important as a means of providing the 'story-line' and of interpreting the entries. They should also, therefore, be explanatory.

Typical paragraphs to follow the entries given as examples above might read along the following lines:

'The constant flow of instructions referred to in items 2.10–2.24 above gave rise to varied and additional works valued at approximately £85,000. This sum represents a 30% increase in the value of the work planned for the period November to the end of February.'

'To appreciate the extent of the delay and disruption which was experienced each instruction should be appraised in the light of the amount of additional work required, the availability of labour and materials and the effect on the other work. In addition to the individual aspects, an examination of the relationship of each item with the others and their cumulative effect on the original programme sequence shows that neither sequence nor project performance could be maintained.'

Specifically:

(a) there was a substantial increase in the volume of work to be executed
(b) deliveries of materials required for additional or varied work were extended

(c) labour resource requirements were subject to fluctuations.

Due to the volume of additional work and the uneconomical work resulting from disruption, work could not be executed in accordance with the programme and the planned labour resources were insufficient

(d) the enforced change of sequence and timing of operations resulted in disruption and uneconomic working

(e) it was necessary for operatives to re-enter areas of the building where work had previously been completed in order to undertake additional works.

'The result of the above items was the need for constant planning, replanning and rescheduling of work and resources to accommodate the work arising from the numerous and frequent instructions and variations. The regular progress of the work could not be maintained and uneconomical working methods had to be instituted to reduce as far as possible delays which would otherwise occur. Productivity was impaired. The financial consequences of these delays, disruptions and actions are set out in the statement of claim.'

Management involvement The delays, variations and additional works, disruption, etc, referred to above inevitably involve management in additional works. It may even have been necessary to employ an additional manager but at the very least the assistant manager, engineer (or whatever he might be titled) responsible for the project in question will have found it necessary to devote additional time, which could otherwise have been spent on other projects, to sorting out the problems arising on the project with which this claim is concerned. The duration of his involvement will also have been longer in this instance.

Much of this involvement will be known to those concerned with the project and obvious from the entries in the statement of events but it must nevertheless be spelt out in the claim. Claims are frequently handled by separate departments when the client is a large organisation and the contractor must have in mind that his claim may, if all else fails, be dealt with by arbitration and an arbitrator can only arbitrate on what is put before him. Furthermore, the contractor will probably wish to put a figure to his additional management costs and he must have provided an indication of the nature and extent of additional management involvement if he is to put forward a sum which carries conviction.

An introductory paragraph under a heading 'Additional involvement of management' might read:

'The sub-contract works were scheduled for execution between ... During the first period the volume and nature of the works were such that the sums contained in the tender proved to be adequate. During the second period, however, considerable additional management was required to deal competently with the substantially increased volume of work and to reduce the delay and disruption caused as a direct result of variations to the planned progress of the contract works.'

'Because of the number, nature, sequence and timing of instructions from the architect which affected the original works, additional management was fully engaged on the following tasks:

(a) attending meetings
(b) preparing and amending schedules, enquiries and orders for contract works, materials and plant
(c) arranging cancellations and postponements in respect of the last
(d) appraising and rearranging labour requirements
(e) planning and replanning and supervising the additional and varied contract works
(f) measuring and valuing variations
(g) budget planning, accounting and financial control.'

Similar details may be prepared in respect of other members of the management team.

Opinions vary regarding the format of claims generally but particularly with regard to the extent to which data is distributed between the statement of events and the statement of claim. On balance, however, it is desirable to extend the statement of events and to make the latter statement as brief as possible.

Statement of claim A statement of claim may, then, read along the following lines:

'From the foregoing statement of events it is apparent that:

(1) The contract period was from . . . to . . . a period of . . . weeks but the extended period was for . . . weeks, an extension of x weeks.

(2) The causes of disruption of the regular progress of the sub-contract works and the extended period were:

    (a) variations (see summary of variations)   x weeks

    (b) disruption (see item . . .)   y weeks
                                              x&y weeks'

Global approach or detailed statement? There is a school of thought which argues that, as there are frequently so many and various interrelated causes and effects of delays and disruptions on building projects, no attempt should be made to attribute costs to individual items. It follows from this that a calculation of loss and expense is made, based on the difference between the cost actually incurred and the sum recommended by the contractor or quantity surveyor as the final account sum.

Cases[1] supporting the global approach are J Crosby and Sons Ltd v Portland UDC (1967) and Stanley Hugh Leach Ltd v London Borough of Merton (1985)
In the first case it was successfully argued on behalf of the contractor that the extra cost incurred depended on the extremely complex interaction between the consequences of various matters such as late possession, suspensions and variations and that it was difficult or even impossible to make an accurate apportionment of the total extra cost between the several causative effects.

The second case also gave support to the global approach.

Some doubt may have been thrown on the method by the decision in *Wharf Properties Ltd and Another v Eric Cumine Associates and Others (1988)* where the Court of Appeal in Hong Kong stuck out the majority of the items in a statement of claim on the grounds that it did not disclose a reasonable cause of action and was otherwise an abuse of the process of the court. It was held that insufficient detail had been given in the claim.

This global approach has the advantage of simplicity and the merit that it is more difficult for the architect or quantity surveyor to whittle away the individual items.

The principal disadvantages are that it lacks credibility and it takes no account of the contractor's inefficiency. It is difficult for a contractor to convince the quantity surveyor or, if need be, the arbitrator that he should reimburse the contractor in full. Similarly, whilst a contractor may be convinced that he is 100% efficient he should not assume that this opinion is automatically shared by all other parties to the contract who may argue: 'Why should we subsidise a contractor's inefficiency?' But whichever approach the contractor adopts, detailed statement or single-sum calculations will have to be made.

No explanation is required of the global calculation which represents the difference between the contractor's prime cost (with or without an addition for profit) and the sum which the client is prepared to pay in respect of 'final account'.

Calculations for a detailed statement of claim are another matter A claim is made for loss and/or expense incurred. The contractor should therefore be able to establish the basis on which he prepared his tender so that he is able to prove that his actual costs have exceeded his estimate. The importance of preparing a tender in a logical, methodical manner is at no time more readily appreciated than when the contractor is attempting to prepare a claim.

With allowances in the tender for, say, management costs as the starting point and with his prime cost book as evidence of actual expenditure the contractor is able to calculate the difference between tender and expenditure.

Management and supervisory costs If an additional manager was fully employed on the project for a period of time the costs in connection with his employment are easily extracted from the prime cost ledger.

The additional cost of managers who would have spent part of their time managing the project in question but who, by virtue of the delays, variations, etc, were more heavily involved, should be charged on a pro rata basis. The statement of events or the preamble to the statement of claim should indicate the tender allowance and the actual involvement: eg

'The tender allowance for the supervisor was one day per week and this proved sufficient during the original contract period. Subsequently in the extended period it became necessary for the supervisor to relinquish his duties on all other projects and devote the whole of his time to this project. The additional involvement was four days per week for the period from 1 July to 14 February, a period of 33 weeks.'

It is important that all such statements are numbered to facilitate reference to them in the calculations. The contractor should be able to provide evidence from his tender of the allowance made for supervision.

The cost of additional management may be calculated using the formula:

$$\frac{\text{Employment cost (£s per annum)}}{\text{Working weeks (or days in year)}} \times \text{additional time (weeks (or days))}.$$

Applying the formula to the supervisor in the example given above and assuming:

Employment cost = £20,000 pa
working year = 46 weeks
working week = 5 days
additional days/week = 4 days
additional weeks = 33 weeks

The calculation would be:

$$\frac{£20,000}{46 \text{ weeks} \times 5 \text{ days}} \times 4 \text{ days} \times 33 \text{ weeks}$$

$$= £11,478.26.$$

When assessing the 'employment cost' of the supervisor, care must be taken to ensure that all costs in connection with his employment are included in the claim.

The above approach may be applied to all staff, including head office staff, if it is not intended to charge the cost of their involvement as part of the 'establishment cost' by means of, say, a final percentage addition. The method of allocating establishment cost which was used when preparing the tender should provide the basis for the claim.

Supervisory visits 'Occasional' supervisory visits during the extended contract period may be calculated in a similar manner to that described above or on a 'cost per visit' basis:

Charge for supervisor's time	
1 day per visit	80.00
Travelling: 150 miles at 35 p/mile	52.50
	132.50
35 visits at £132.50	£4,637.50.

Labour and plant costs The labour costs of some specific 'extra' items or work may be clearly identifiable in which event they may be charged as such. In some instances, however, labour costs may have increased due to wage awards, non-productive overtime and travelling time (due to the need for men to work longer hours or travel from greater distances, etc, than was anticipated at the time of tender).

Such additional costs may be calculated for the extended contract period and listed in detail by showing the hours worked at the increased rate.

Plant costs are valued in a similar manner to labour.

Non-productive overtime This may be calculated from time sheets making allowance for any provision for such time as had been included in the tender.

Loss of productivity Loss of productivity may be calculated by assessing the ratio of labour to materials in the tender and comparing it with the ratio in the work as executed. If the tender was prepared by building up the tender sum by adding together the sums for labour, plant and materials, in a tender summary the sums included are easily ascertained.

Assume, for example, the tender summary indicated that the sums included were:

   Labour costs £380,000
   Materials costs £500,000
   giving a ratio of labour : materials
     = 43.18 : 56.82.

If the prime cost ledger at completion of contract records:

   Labour costs £750,000
   Materials costs £630,000 giving a ratio of 54.38 : 46.65.

The higher labour ratio may be related to the estimated labour cost included in the tender. It will be seen that a loss figure on the original work value may be calculated as:

$$\frac{54.38}{43.18} \times £380{,}000 - £380{,}000$$

$$= £98{,}564.20.$$

The calculated sum relates to the tender sum and it does not take into account any variations which may have occurred during the course of the contract works. These should be dealt with separately.

It will be appreciated that the ratios based on the prime cost ledger at the completion of the contract must be regarded as merely indicative. They do, however, provide a simple basis of calculation which, in the absence of a better method, enable the contractor to demonstrate that he is at loss and that his records show a different picture from that which he anticipated at the time of tender and that which he contracted to carry out and complete.

Clearly, too, the ratios of labour : materials have been changed from those on which his tender was calculated.

Labour is usually the most difficult ingredient of work to estimate, and to monitor and control during the course of the works.

It follows that the majority of a contractor's losses are incurred through his failure to achieve levels of productivity which he considered to be reasonable at the time of his tender. This failure is frequently due to circumstances beyond his control.

Materials, plant and 'establishment' are, however, other causes of loss.

If labour may be described as the most variable ingredient it is true to describe material as the most visible ingredient – material is the product itself.

Material costs The material content of claims is frequently less subject to controversy than the other ingredients because of its visibility.

Comparison of tender with cost When assessing the extra cost of materials the starting points should be the contractor's tender and his prime cost ledger. Comparison of the sums of these items will show loss or profit. If the prime cost ledger sum does not exceed the sum in the tender summary no action may be necessary in regard to the claim but if the prime cost sum exceeds the tender sum the contractor must look further.

Materials abstract An abstract of material contained in the prime cost ledger should be prepared which identifies the quantities of all items. This is most conveniently prepared on specially ruled abstract paper with sufficient space allowed on the sheet between items to make it possible for 'credits' to be entered for the materials for which the contractor has been paid in the final account. The contractor is paid for the materials in the final account under the headings of 'contract sum' or 'variations'.

The quantities of materials to credit in the abstract against those contained in the prime cost ledger are, therefore, taken from the contractor's original estimate plus those contained in variations. Materials in variations which give rise to omissions are regarded as 'debits'.

A comparison of the number of any units purchased with the number 'built in' should show a balance but if, for example, more have been purchased than built in one should ascertain what caused the difference.

Possible causes are:

(a) estimating error. The estimator failed to measure the full extent of the work

shown on the drawings. The contractor will have to bear the loss (but profit from the experience)
(b) bills of quantities error. Similar to (a) but in this instance it is assumed that the contractor tendered on the basis of a bill of quantities prepared by a quantity surveyor. The contractor is entitled to payment in this event but he may have to measure the works as shown on the original drawings before the quantity surveyor will be convinced of the error/s
(c) defective workmanship. Items of work incorrectly positioned, damaged during fixing, etc, will be made good at the contractor's expense
(d) theft or loss. Another contractual risk against which prevention is the best safeguard
(e) additional works carried out in accordance with architect's instructions or with contractor's directions in the case of specialist sub-contractors but which have not been accounted for as variations. The contractor should check instructions against variations
(f) waste due to breakages caused by the contractor or other contractors after installation by the contractor.

The important factor for the contractor is to identify the difference. It is then possible for him to ascertain the causes and submit a claim for items for which he may have a contractual entitlement.

Cost differences In addition to variations in quantity, there may be variations in purchasing costs which will be identified in the abstract. Possible causes are:
(a) inflationary variations since entering the contract
(b) variations in rates due to purchasing in small quantities
(c) suppliers' quotation errors
(d) invoicing errors.

NEDO formula When there is provision in the contract for automatic cost adjustment by means of a schedule of basic rates or the NEDO formula, as occurs with 'fluctuation' contracts, the contractor should have less difficulty in obtaining reimbursement than might be the case on 'firm price' contracts. On firm price contracts the contractor may consider the use of the NEDO formula indices as a way of demonstrating the extent of his loss during an extended contract period. The effect of inflation, cause (a) above, frequently represents the major cause of a contractor's loss. Increases resulting from small quantity purchasing rates, cause (b) above, should be recovered as variations if the materials were purchased in small quantities as a result of architect's instructions or contractor's directions.

Claims against sub-contractors, suppliers and insurers The contractor may decide to claim against his sub-contractors and are his suppliers for any loss which he may incur in respect of quotation errors and/or invoicing errors (cause (c) and (d) above). He may have a claim under an insurance policy in respect of courses (d) or (f).

Site supervision and management costs Provided an estimate has been prepared in a systematic manner it should be possible to identify the provision made for site supervision and management costs.

For purposes of illustration assume that these costs were estimated at £75,000. This sum would be derived from the contractor's estimate of all the costs which he would incur in providing staff, plant, storage sheds, offices, telephone, etc, in order to run the project. If he has used a checklist or standard proforma for this purpose it will be a simple matter for him to ascertain the cost of individual items included in his tender and thus calculate the additional cost which he has incurred as a result of the extended contract period, etc. He will also, of course, be able to justify his claim by reference to his estimate, if he is called upon to do so.

Additional costs of providing offices, storage sheds, furniture, plant (such as benders, steps, ladders, extension leads, welding equipment, scaffold towers, etc) may be assessed on a hire cost per week basis and the total cost per week may be multiplied by the number of additional weeks due to the extended contract period.

The matter of site staff, supervisors and managers were discussed above when methods of calculating additional costs were explained but the contractor may prefer to calculate the additional cost of site management using the formula:

$$= \frac{\text{site management cost} \times \text{\# no weeks in extension}}{\text{weeks in original period}}$$

$$\frac{£75,000 \times (\text{say}) \ 15 \ \text{weeks} - £75,000}{(\text{say}) \ 10 \ \text{weeks}}$$

$$= £105,000.$$

Quantity surveyors may be less inclined to accept such 'single sum' adjustments but in the absence of a detailed build-up of the contractor's estimate there may be no alternative method of calculation. The contractor should experiment with alternative methods of calculation to ascertain the one which is to his best advantage. If, however, he decides on the single sum approach he should bear in mind that the quantity surveyor might well suggest that some items in his estimate would not be affected by an extension of time and disallow part of the sum. The cost of erecting and dismantling temporary offices and sheds are examples of items not affected by an extension of time. Whilst site management costs are normally time-based, being directly related to the duration of the project, some contractors make provision for these costs as percentage additions.

Establishment/head Office (overhead) costs

There is legal precedence for the payment of a contractor's establishment costs as a head of damage if he can offer sufficient evidence in support of the costs to satisfy a judge or arbitrator.

An auditor's statement that the contractor's establishment costs for former years were x% might be considered sufficient evidence. Alternatively, an authentic tender analysis showing the amount included by the contractor in his tender for establishment costs and profit might be used as evidence.

Formulae At least three formulae are used for the calculation of establishment costs, best known of which is probably the Hudson formula described in *Hudson's Building Engineering Contracts*.

The Hudson's formula is expressed as:

$$\frac{h}{100} \times \frac{c}{cpw} \times pdw$$

where

h = head office overheads and profit included in the contract (as percentage)
c = contracts sum (£s)
cpw = contract period in weeks
pdw = period of delay in weeks.

Emden's formula comes from another book on contract law, *Emden's Building Contract Practice*.

It may be expressed as:

$$\frac{h}{ct} \times 100$$

where

h = total of company head office overheads and profit (£s)
ct = company turnover (£s).

The third alternative is Eichleay's formula which is developed in three stages:

STAGE 1

$$\frac{fa}{ct} \times hp = hac$$

where

fa = final account (referred to as 'contract billing' in original formula) (£s)
ct = company turnover (£s)
ct = head office overhead during contract period (£s)
hac = head office overhead allocatable to contract (£s).

STAGE 2

$$\frac{hac}{lappd} = had$$

where

hac = head office overhead allocatable to contract (£s)
appd = actual days of contract performance (actual project period in days)
had = head office overhead allocatable to contract per day (£s).

STAGE 3
had = ndc = ar

where

had = portion of head office overhead allocatable to contract per day (£s)
ndc = number of days of compensatable delay
ar = amount to be recovered (£s)

It will be seen that the Hudson formula requires less data about the company's finances than the other two. Armed with all the data which are available one is able to apply the formula to determine the most appropriate solution.

Turner (see *Further reading*) suggests that running the formula into one and using British terminology, the amount to be recovered may be expressed as:

$$\frac{\text{amount to be}}{\text{recovered}} + \frac{\text{total overheads}}{\text{for contract period}}$$

$$\times \frac{\text{final account sum}}{\text{total accounts}} \times \frac{\text{days of delay}}{\text{days of performance}}.$$

Whichever formula is adopted the contractor (or sub-contractor) will have to prove the following in order to succeed with his claim for reimbursement of establishment cost and profit:

- that he has a sustained ability to make profit, referred to above
- that there is a sustained market for his services
- that there is a demonstrable basis for the calculation of the percentage included for his establishment cost
- that as a result of the project in question he could not undertake further work in order to mitigate the loss.

Cases relating to establishment costs[2] include *Ellis-Don Ltd v The Parking Authority of Toronto (1978) 28 BLR 98* Here the plaintiffs had been delayed for some 32 weeks on a car park contract. It was found that 17.5 weeks of the delay was due to the employer's failure to obtain an excavation permit and the contractor had been thrown into winter working as a result. The High Court of Ontario ruled that the contractors were entitled to recover as damages for breach of contract: (a) the extra cost of pouring concrete in winter and (b) the on-site cost of 17.5 weeks. A weekly sum of overheads and loss of profit calculated by reference to 3.87% had been built into the tender which was assessed in essence by reference to the Hudson formula. The court accepted that there was no logical distinction to be drawn between a claim for lost profit and a claim for lost head office overheads and appeared to accept the Hudson approach.

J F Finnegan Ltd v Sheffield City Council This case involved a claim for damages for prolongation and loss of productivity which included head office overheads. It was held that the contractors were entitled under clause 24 of JCT 63 to 'direct loss and/or expense' for prolongation of work by reasons of the matters set out in the clause. Sir William Stubb QC, went on to consider the question: 'What if any allowance for overheads and profit should be made on the plaintiff's claim?'

He said:

'It is generally accepted that in principle, a contractor who is delayed in completing a contract due to the default of his employer may properly have a claim for head office or off-site overheads during the period of the delay, on the basis that the workforce, but for the delay, might have had the opportunity of being employed on another contract which would have had the effect of finding the overheads during the over run period.'

This principle was approved in the Canadian case of *Shaw and Horowitz Construction v Frank of Canada (1967) SCC page 589*, and was also applied by Recorder Percival in the case of *Whittal Builders Company Ltd v Chester-le-Street DC*.

Tate & Lyle v GLC (1982) 1 WLR 149 This is not a building case but the principles are important. In this case the plaintiffs had claimed 2.5% for managerial time and the judge accepted that such a head was acceptable but he did not accept the method of calculation and the application was rejected. He remarked that it was up to the manager to keep time records of their activities. A different result may have been achieved had the claimed 2.5% been authenticated, ie an auditor's certificate.

Wraight v PHAT (Holdings) Ltd 1968 13 BLR 26 The case concerned the construction of the determination provision contained in clause 26(i) (c) (iv) of JCT 1963, where the expression used is 'direct loss and/or damage'. It is widely thought that the two words mean the same as 'direct loss and/or expense'. The two main issues decided in the Wraight case were:

– the words 'direct loss and/or damage' mean the same thing as the amount of damages that would be recoverable under ordinary legal principles for breach of contract
– accordingly loss of gross profit is 'direct loss and/or damage'.

In this context, gross profit equates to the amount a contract should have contributed to the cost to the contractor of maintaining his head office establishment and also the anticipated contribution towards profit.

Finance charges Cases which support claims for reimbursement of loss and expense involving financing the works are *F G Minter Ltd v Welsh Health Technical Services Organisation* (1980) and *Rees & Kirby Ltd v The Council of Swansea* (1985) but proper notice in advance must be given if the contractor is to succeed in his claim. Subject to notice having been given the contractor has an entitlement to be reimbursed for the loss and/or expense he has incurred as a result of paying interest to the bank, finance company, etc, on cash borrowed to finance works which are themselves the subject of an application for reimbursement because the regular progress of the works has been disturbed.

The contract makes provision for payment on account and the contractor should not suffer loss as a result of the extension of the contract period, disruption, etc.

The contractor's claim might include for reimbursement of financing costs:

(a) additional work whilst in progress
(b) additional retention held during the original contract period and retention held during the extended contract periods
(c) work in progress during the extended contract period
(d) after practical completion
(e) the settlement period.

Figure 2.3, shows a graph with expenditure profiles. The 'budgeted cost curve' shows the contractor's planned expenditure based on his estimate. It takes into account the original contract period of 12 months.

The 'actual cost curve' shows the value of work executed over the extended contract period of 16 months. These values may be derived from the contractor's prime cost records or valuations based on the programme.

The additional financing costs incurred by the contractor may be based on the difference between the two curves.

Profit The tender summary should show the contractor's planned profit as a percentage on his estimated cost of labour, plant, materials, etc, or as a lump sum. The allowance made in the estimate should provide a basis for calculating the appropriate allowance for profit in a claim for reimbursement.

Cost of preparing the claim This may be regarded as a special part of the contractor's establishment cost and charged accordingly. These costs may be more readily demonstrated if the claim has been prepared by a consultant who has charged a fee for his services.

Main contractor's discount Nominated sub-contractors are required to allow the main contractor 2.5% discount and one-thirty-ninth should be added to the total.

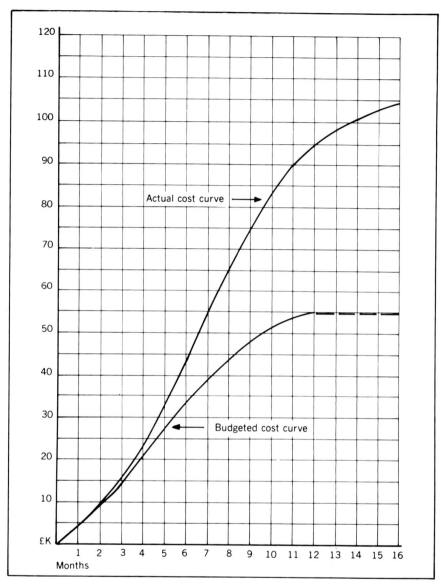

2.3 *Expenditure profiles*

Presentation of claim The 'packaging' of claims is not an aspect which is taken very seriously in the building industry, so the contractor need not indulge in sophisticated artwork but he may find that photographs and illustrative diagrams are useful to make his case: 'a picture is worth a thousand words'. He should also provide detailed, supportive information in the form of appendices so that the story-line of his claim is not disrupted by unnecessary detail.

The contractor may decide to deliver the claim personally – the personal touch has much to be said for it – and it may be expedient to follow up by telephone after a suitable lapse of time. The 'first come first served' principle applies as much to claims as to other business matters. Following up is also useful as a way of opening the way for negotiation. An enquiry regarding points on which the client may require clarification or may require further information can be used as the reason for the contractor making the call.

Few claims are accepted in their entirety and a meeting between the parties may provide an occasion for negotiating on some items.

No two clients are the same so the level of acceptance of the claim will vary from client to client. The primary rule for negotiation is to avoid placing the other negotiator (and oneself) in an irreversible position. A first meeting might be planned with a view to sounding out the opposition rather than reaching firm agreement on cost matters.

The contractor should, as suggested above, have provided his claim with some spare capacity. Whilst negotiations should not be conducted on the lines of an oriental marketplace no negotiator would wish to report that he had spent time examining documents and attending meetings without, at least, obtaining a reduction in the claim in excess of the cost of the time he devoted to investigating it.

Unless the contractor is unusually successful in his negotiations it is unlikely that his claim will be accepted in its entirety and he may have to decide wether or not to accept the sum which his client is prepared to pay. At what point should the negotiator 'stick'?

Logic, even common sense, is seldom able to compete with emotion when it comes to negotiations and acceptance of a 'final' offer but the economic fact of claims is that a balance must be struck between the offer which has been made to the contractor, the possible award which an arbitrator (if the matters goes as far as arbitration) might make, the cash-flow embarrassment which the contractor might experience whilst awaiting the arbitrator's award, and the cost of legal and arbitration fees if the contractor's decision is to decline the final offer and refer the matter to arbitration.

As a ploy, a movement in direction of arbitration might lead to an improved final offer, but reference to arbitration by the contractor is a serious decision which should not be made unless he is very sure of his grounds. Occasionally, as a matter of principle, arbitration may appear to be the only solution but principles can prove to be expensive luxuries.

Who prepares the claim? The contractor must decide who will prepare the claim. Will it be his own staff or a specialist in such matters?

The larger contractor may well have an accounts or quantity surveying department which has personnel experienced and qualified in the claims field but smaller firms frequently do not have such departments and they must decide whether to arrange for their own staff to find time to take aboard the considerable work commitment involved in claim preparation or to engage a specialist consultant.

What are the advantages and disadvantages of the alternative courses of action? These can be considered under several headings.

Available staff time Unless the time of the staff who normally carry out estimating, surveying and the technical accounting functions within the contractor's own firm is less than fully committed the staff may find it necessary to prepare the claim as an 'overtime' task. This may mean that processing the claim will be a prolonged task and that monies which the contractor has outstanding will not be recouped in the near future. Furthermore, the contractor may find a deterioration in his relationships with his client if the matter is not dealt with expeditiously. One of the banes of building contractors is the time which some sub-contractors take to submit their accounts. When considering contractors for future projects, performance with regard to accounting matters will certainly be considered.

Staff knowledge and experience The preparation of claims is not a task which, hopefully, a contractor will be required to undertake on a regular basis and he will probably lack the necessary knowledge and experience for the task. The ingredients of a claim are very important. The contractor should be aware of the necessary 'input' but the recipients of a claim may take greater note of a submission which appears 'professional' than one which is home-spun. This will depend to a considerable extent on the client's degree of sympathy with the contractor's cause. If the client has made it known that he is sympathetically disposed to a claim but that he needs dates, facts and figures before he would feel justified in recommending payment a sophisticated submission might be unnecessary and, indeed, even counter-productive.

If, on the other hand, the client is likely to resist a claim and the contractor anticipates the need to resort to arbitration, a professional submission may be of considerable importance.

Cost of preparation is high regardless of whether the contractor's own staff or a consultant undertakes the work. If, however, the contractor's staff are working at less than their potential capacity, in-house preparation may be more economical provided the expertise is comparable with that which should be obtained from a consultant.

A consultant will often give an estimate of his fee for making an initial examination of the papers and advising the contractor regarding his case. Having ascertained the strength of the case, he may offer to prepare the claim, on a 'percentage of costs recovered', on a daily rate or on a 'lump-sum fee' basis.

The 'percentage of costs recovered' basis is probably the most positive for the contractor but consultants will not usually work on this basis unless they consider the contractor's case to be good and the sums involved are sufficiently high.

A 'lump-sum fee' estimate ensures that the contractor knows the extent of his commitment but consultants may pitch their estimates high because it is difficult for them to assess the extent of their work in advance.

Whilst most consultants will wish to take over the whole of the papers and do all the work within their own offices and employing their own staff a few will work with the contractor's staff in an advisory capacity. In this event the contractor undertakes much of the routine 'searching' with his own staff and assembles the information under the guidance of the consultant who is then able to interpret it and draft the various statements referred to earlier in this section. The largely routine clerical work, photocopying and assembly can usually be accommodated by the contractor without undue disruption of his staff's regular work and at less cost than would be possible if the consultant used his own staff. Furthermore, the contractor may obtain experience whilst working with the consultant which will be useful for the future.

The approach to preparing a claim which has been discussed above is not as detailed as that adopted by some consultants but it should provide an indication of many items which he should take into account when seeking reimbursement.

Credibility of claims It must be remembered that on very few occasions will the picture presented by the contract records be painted in just black and white. There will be shades of grey where the architect, quantity surveyor, contractor or, perhaps, arbitrator may have to exercise his discretion when considering the recommendation he will make to the client regarding the payment which should be made to the contractor.

The human factor enters into the question of claims and it is, therefore, important that the contractor's claim appears to be credible. The contractor should concentrate on clarity and simplicity of presentation. Architects and quantity surveyors are sometimes suspicious of claims which appear to be 'clever' and to react against them. There is much to be said for the contractor having the person to whom he is submitting the claim on his side.

2.4 Role of the expert witness

Passing reference is made in section 4 to the role of the expert witness in connection with the resolution of disputes. Experts are frequently engaged before a dispute arises to provide objective, independent opinions on the strengths and/or weaknesses of their client's 'case' in the event of the progress of the works being disrupted and of there being the prospect of loss. The expert's task being initially to safeguard the client's interests and avoid a difference or dispute developing. If the dispute develops he may find himself called as an expert witness.

Practitioners from all disciplines in the construction industry may be engaged as expert witnesses. They may stand as experts regarding technological matters, custom, usage, contractual practice or cost and value. Experts are frequently engaged to give evidence regarding loss and expense claims. A practitioner with appropriate expertise and experience who is an employee of one of the parties to a dispute may act as expert but his independence might be questioned by the opposition or by the court.

This section is concerned primarily with the role of the expert in matters of contractual procedures, delay, extension of time, loss and expense and application for reimbursement, etc.

Briefing The expert obtains his instructions from his client or his client's solicitor. 'Client', in this context may mean any of the parties to the contract; typically the employer, contractor or sub-contractor.

The expert may find it advisable to take a checklist with him to the briefing to ensure that he records the essential facts about the case. Items such as the names of the parties, nature of the documentation, sources of information regarding fact, etc.

At the conclusion of the briefing (or soon after), both parties should agree the expert's terms of reference. These should be precise and unambiguous.

Evidence The expert should study all the evidence, read all relevant documentation and speak to all concerned. He should note all relevant facts and figures in so far as required by his terms of reference.

Documentary evidence available to the expert is referred to in section 4.6. Oral evidence may be obtained from all concerned with the project including estimators, contract surveyors and managers, site personnel and head office staff.

Preliminary report The principal aim of the preliminary (or initial) report is to indicate the client's 'prospects' in regard to the case. The expert should emphasise the stronger arguments which favour the client but he should identify arguments which do not.

The report must be independent, impartial and balanced. If the expert receives assistance from others with the preparation of his reports their contributions must be credible to the expert.

The report will almost certainly be the subject of a preliminary meeting of the client, his legal advisers and other members of the client's team. At such a gathering of 'like-minds' it is tempting to over-emphasise matters which favour the client. This is a trait demonstrated at political party conferences.

There is a risk of 'group think' developing.[3] The expert must remember when encouraged to make a strong case during such a meeting that should the case go to court he may find himself facing aggressive cross examination. The witness stand is a lonely place.

Format of report This should:

- set out the terms of reference
- address all the issues in the terms of reference giving information and explanations
- have an introduction, body, conclusion and summary
- be drafted with a view to it being read by a judge
- be factual, concise and to the point
- follow a logical path, use simple words and short sentences
- use 'lay' terms and avoid jargon

- number the paragraphs for ease of reference (the JCT 80 and similar conditions of contract provide models in this respect)
- allow space in the form of wide margins or between paragraphs for the reader's annotations
- include all calculations necessary for the report
- be able to stand alone, to be complete in itself and capable of being read without reference to other documents
- include appendices for matters necessary for the report but which would interrupt the flow if included in the body of the report. (Calculations might be included as appendices.)

The report is proof of the expert's evidence and is referred to as 'proof of evidence' in the court.

Consultation with other experts The expert should discuss his preliminary report with other experts in the 'team' to avoid discrepancies between or conflict with other experts' reports but his report should retain its integrity.

The finalised report should be submitted to the client (or his solicitor).

Exchange of reports Experts' reports are frequently exchanged with those of the 'opposition'. Opposition reports should be studied critically. Their strengths and weaknesses should be noted and reported.

Meeting of experts Reference is made in section 4 to meetings of experts which are held without prejudice. Their aim is to explore common ground and endeavour to reduce the differences between the opposing parties or even to reaching settlement.

These meetings should be conducted to agenda agreed between the experts in advance of the meeting.

The pre-trial meeting is attended by members of the team who will be involved in the trial. The aims of this meeting are:

- to discuss arrangements for the trial
- to discuss tactics to be adopted
- to examine the strengths and weaknesses of both sides
- to familiarise members of the team with facts and figures (there are often considerable lapses of time between preparation of evidence, meetings of experts and the hearing).

Trial/hearing During the trial the expert witness should:

- stay within his area of expertise (he will lose his credibility if he strays)
- attend as much of the trial as possible
- advise counsel on the evidence given by opposition experts
- advise on cost and value
- endeavour to be accessible at short notice if not present in court (stay near a telephone during the trial)
- respond to counsel (of both parties) and to judge/arbitrator
- make his responses during cross examination as concise as possible
- disagree with the evidence of opposition, if appropriate
- maintain detailed diary of his time and activities and accounts of disbursements as basis for his fee (detailed accounts may be requested for 'taxation' purposes if the amount is considered to be excessive).

Expert's integrity The role of expert witness requires considerable professional and ethical integrity. The aim of cross examination is to discredit the witness's evidence from which it follows that the expert puts his reputation at risk when he offers his services as a witness. It is not a role to be undertaken lightly.

2.5 Case studies

STUDY 1 *Hawthorne Home Scenario*
The scenario contained in section 1.12 provides material for study in section 2. Some exercises in section 2.6 relate to the Hawthorne Home scenario.

STUDY 2 *Tender, cost and value analysis for Alpha Industrial development*

The following particulars provide further material for study in section 2 generally and for exercises in section 2.6

The particulars comprise:

1 *Tender analysis* under the principal headings. Item A includes the cost of labour, plant and materials and works which will be undertaken by sub-contractors. Item D is based on previous years' audited accounts. Item E is the margin added for profit. The contract period is self-explanatory.

2 *Actual (historical) costs* Items A1 and B1 are taken from the prime cost ledger. Items D1 and E1 adopt the same percentages as those used for the tender. The works took 35 weeks to complete. An extension of time for 2 weeks has been certified by the architect. The employer has claimed and deducted liquidated damages (as stated in the appendix) at £10,000 per week for the 3 weeks not included in the extension of time. The liquidated damages deducted of £30,000 are not taken into account under headings 1 to 3.

3 *Final account* This is the draft proposed by the consultant quantity surveyor but not yet agreed with the contractor.

4 *Company performance* The figures are taken from the audited company accounts except for the closing item which is based on calculations made by the company's financial manager.

1 *Tender analysis* $
  A   Works   980,000
  B   Preliminaries   85,000
  C   1,065,000
      Establishment/profit and risk comprising:
  D   Establishment charge 4%)
  E   Profit and risk 6%) 10%   106,500
  F   Tender sum
                                       TOTAL 1,171,500

Contract period 30 (5 day) weeks

2 *Actual (historical) costs*
  A1   Works   1,450,000
  B1   105,000
  C1   1,555,000
  D1 & E1   Establishment, profit and risk   155,500
  F1   Prime cost
                                       TOTAL 1,710,500

Time taken 35 weeks

3 *Final account*
  A2   Works, including variations, etc 1,050,000
  B2   Preliminaries   90,000
  C2   1,140,000
  D2 & E2   Establishment, profit and risk   114,000
  F2   QS's proposed final account
                                       TOTAL 1,254,000

4 *Company performance*

	Turnover £K	Establishment cost £K	%	Profit (loss) £K	%
3 years ago	30,000	1,800	6	1,500	5
2 years ago[1]	35,000	1,750	5	700	2
Last yr (yr of tender)[2]	40,000	1,600	4	2,400	6
This year	50,000	1,500	3	3,500	7
Next yr (projected using work in progress projections)	70,000	2,100	3	4,900	7

[1] Losses on major projects
[2] Corporate reorganisation.

2.6 Exercises

1 Situations often arise where a contractor is able to pursue a contractual claim against the client. *Discuss* how a claim situation may be identified and indicate the records which should be kept to ensure maximum financial recovery

2 A substantial and complex design variation is introduced to a building project being carried out under JCT 80.
   (a) *Consider* the problems which may be created by this variation and explain the means whereby the contractor may secure adequate financial reimbursement
   (b) *Describe* the documentation required to aid evaluation and offer validity to any contractual claim.

3 A two storey office block of traditional construction is being built under JCT 80.
   While work is proceeding on the first lift of brickwork, the architect issues an instruction requiring the prestressed concrete first floor to be omitted and a reinforced concrete floor to be cast in situ in accordance with the structural engineer's design.
   Discuss the rights of the contractor and the actions he can take under the contract to maintain his financial position. (Clause numbers are to be quoted where appropriate.)

4 An architect's instruction requiring a variation may have serious effects upon the work in respect of both disruption and resultant cost implications.
   Comment upon this statement and critically analyse the adequacy of the contractor's rights under JCT 80.

5 On a contract using JCT 80, a nominated sub-contractor fails to progress the work in accordance with the master programme of works and this causes a delay.
   (a) *Explain* how the contractor can recover the loss and expense resulting from this delay.
   (b) *Explain* how the contractor can ensure that he will not have to meet a claim for liquidated and ascertained damages from the employer arising out of this delay

6 A project is let under the JCT 80. During the course of excavating for new drainage, the contractor damages an electric cable causing a loss of power to his own tower crane and to an adjacent amusement arcade. Two days elapse before repairs can be effected. The cable was not indicated on the contract drawings.
   Discuss:
   (a) the likely additional payments which the contractor might be successful in obtaining from the client
   (b) the insurance position.

7 Adopting the Hawthorne Home scenario in section 1.12 and assuming the role of contractor in week 13 *draft* a letter to the architect which takes account of the situations which occurred in weeks 4, 5 and 8.
   It is appreciated that you might not wish to include the appropriate clause numbers in your letter but note them in the exercise.

8 Using the data contained in case study 2, section 2.5, and the three formulae discussed in section 2.3, *calculate* establishment costs for the Alpha Industrial development.

2.7 Sources and further reading

Relevant documents, forms of contract, general texts, etc
See 1.14

Claims and disputes
TURNER D, *Building contract disputes*, Longman 1989 (comprehensive case studies)

Expert witness and evidence
REYNOLDS M P and KING P S, *The expert witness and his evidence*, BSP Professional Books 1988

2.8 References

[1] KNOWLES R, 'Global Claims', *Chartered Quantity Surveyor*, December 1989 contains a more detailed account of global claims
[2] GOLLEY O M and LOMAS CLARKE P *ASI Journal*, September 1989

[3] Group think is discussed in KAST R, and ROSENZWEIG J (1986) chapter 17. In a gathering of like minds, individuals are loath to stand outside the group and fail to express their doubts because of the apparent unanimity of the others.

Section 3
INSURANCE
by Brian Thornton

This section is concerned with:
- explaining the general purpose governing insurance
- providing sufficient information to enable the contractor to know what in general is insurable
- providing information with regard to what he may be required to insure under the terms of a contract or at law
- explaining whose responsibility it is to insure in different circumstances
- explaining whose responsibility it is to endeavour to recover claims money from insurers, and how to go about it.

3.1 Background

An understanding of what can be covered by insurance and what risk a normal contract will require the contractor to insure is essential.

It is also necessary to know when it is sensible to retain a risk as being self-insured, and when it is sensible to transfer that risk to an insurance carrier, ie, an insurance company. Any such decision has to be made bearing in mind the requirements of an individual contract.

The contractor should also know how and when to make a claim against an insurer and various other actions which should or should not be taken regarding arranging insurance or making claims thereunder.

3.2 Principles

Utmost good faith A contract of insurance is said to be a contract of 'uberrima fides' which means 'utmost good faith'. It differs from most commercial contracts which are said to be contracts of caveat emptor, meaning 'let the buyer beware'. For example, subject to no other contractual agreement, a person purchasing a motor car has the right to inspect the vehicle, and as a consequence if he fails to do so or fails to do so adequately it is the buyer's problem should the vehicle not be in the condition that he thought it to be.

In the case of insurance, however, there is no tangible product to inspect, and the purchaser of insurance relies upon the good faith of the insurer to provide the product he requires, but at the same time and probably of far more importance, the insurer relies upon the risk being presented to him fully and adequately without misrepresentation, and therefore in turn relies upon the good faith of the proposer.

In view of the above should a risk be misrepresented to insurers they would have every right to void the policy from the inception date and therefore to decline to pay any claim irrespective of whether that misrepresentation was material to the claim being made.

Proximate cause For any claim to be recoverable under an insurance policy that particular loss must have been directly caused by an insured peril. Insurers, therefore, will look to find out what was the proximate cause of any loss. Proximate cause has been defined as 'the active efficient cause that sets in motion a train of events which brings about a result without the intervention of any force started and working actively from a new and independent source'. Thus a policy covering a building against fire risks only, which was burnt down by rioters, would not pay for the loss as the proximate cause would be deemed to be riot and not fire.

Insurable interest To be able to make a successful insurance claim the claimant needs to be able to prove that he has an insurable interest in the subject matter of that insurance and is therefore likely to suffer a loss in the event of some misfortune occurring. For example, he could not successfully arrange an insurance policy against the risk of fire to his neighbour's house as he does not stand to suffer any loss in the event of it being damaged by fire. However, if he had agreed to take into his house some of his neighbour's property for safekeeping, he may well have an insurable interest in that property in that in the event of it being damaged he may have breached the duty of care, to his neighbour.

3.3 Main types of insurance

Basically, insurance falls into three main categories being in respect of:

1. material loss or damage to property
2. liability which some person or company may incur which may in turn be in respect of damage to property or loss of life, etc
3. life insurance or personal accident insurance. These fall outside the terms of this section, and are in any case subject to slightly different insurance rules.

Practice In this section we will only refer to insurance of loss or damage to material property and insurance of liabilities, concentrating on their application to the construction industry.

3.4 Loss or damage to material property

As will have been seen from the comments on insurable interest, above, a contractor or sub-contractor can only insure against material property in which he has an insurable interest. In a simple arrangement of building contracts, therefore, it is fairly normal for the employer to request that the main contractor insures his own contract works, and the employer often specifies against which insurance perils he wishes to insure. (It should be pointed out, however, that depending upon the terms of the contract the main contractor may well be responsible for all material damage irrespective of what he is actually required to insure.)

The main contractor in turn then sub-contracts out parts of the work and practice varies between contracts as to responsibility for insurance. Some contracts will pass down to the sub-contractor responsibility for all loss or damage to his own works whilst others will either make the main contractor responsible for all loss or damage or may make the main contractor responsible for only certain perils with the balance of the perils being the responsibility of the sub-contractor. Contracts also vary as to whether there is a responsibility to insure those risks on behalf of the sub-contractor, but generally speaking where main contractor (or it may be the employer) accepts responsibility for certain perils there is often an obligation that they should be insured.

It is important, therefore, to look at any contract to establish who has responsibility for the work or sub-contract works as the case may be but, in addition, who has responsibility to insure and for whose benefit. The employer normally insists upon insurance to protect his future assets in his building, and in some cases may also take the responsibility of taking out the insurance himself. However, it is more normal for the employer to request the contractor to insure, and in turn for the contractor to pass at least some responsibility on to the sub-contractor and this may be either with or without any obligation to insure.

Similarly, one must look at the contract to find out what type of insurance is to be arranged, eg, what perils are to be covered. Until 1986 it was normal for insurance to be called for in respect of what are commonly known as Fire and Special Perils. These perils were nominated under a JCT contract, for example, as fire, storm, tempest, flood, lightning, explosion, bursting or overflowing of water tanks, apparatus or pipes, earthquake, aircraft and other aerial devices, or articles dropped therefrom, riot and civil commotion. In later contracts the main contractor is often required to insure against 'all risks of loss or damage' subject to certain acceptable restrictions. Conversely some contracts make no reference to the contractor needing to insure, eg G.C.Works 1, but this does not mean that the contractor is not responsible for that loss or damage and he would be very unwise not to arrange insurance.

The policy issued in respect of such cover is known by various titles. For example, *All Risks Contract Works, Contractors All Risks, Contractors Indemnity* and others. Cover under such policies normally extends to not only the works, but also plant for use in connection with construction, temporary buildings and other items brought on to site for use in connection with the contract. Such a policy does not normally include cover against loss or damage to any buildings or contents which existed on site prior to the commencement of the works. This is often an area of misunderstanding both between contractors and their insurers, and also between contractors and the principal.

In any alteration, extension, refurbishment or similar contract in all normal circumstances the existing property is the property of the employer, and stands at his risk unless the contractor causes negligent damage thereto (see legal liability comments below) and it is for the principal to insure. Attempts are often made to put the responsibility for all loss or damage and insuring that loss or damage on to the contractor via the contract, and if this happens the contractor must make special arrangements with his insurers, if, indeed they will give such cover.

3.5 Legal liability

Insurance in respect of legal liability falls into the following categories:

Employer's liability is insurance by an employer against his liability for any death, injury, illness or disease which may be incurred by any of his employees, and which may occur as a result of that person's employment. 'Employer' under this heading refers to the contractor. The current situation in the United Kingdom is that, with a few exceptions, any person or firm employing one or more persons is obliged to carry Employer's Liability insurance by the Employer's Liability Compulsion Insurance Act of 1969. The construction industry, regrettably, does not have a good record in respect of injury to employees and consequently the premiums charged by insurers tend to be fairly expensive. The claims history of the individual person or company taking out that insurance affects the premium charged.

This policy covers any liability of the employer in respect of death, injury, illness or disease, generally without any limit on the amount of such claim. There are, however, a number of insurers who tend to impose restrictions by way of exclusion of certain types of work and/or methods of work, and if an employee is injured as a result of his employer carrying out work in an excluded category an interesting situation arises. The existence of an Employer's Liability Compulsory Insurance Act will normally lead insurers to handle the claim on behalf of the employer thereby making certain that in the event of negligence by the employer the injured employee is paid. The insurer then invokes the right under the terms of the policy to recover any money paid from the employer. This is done to make certain that the injured party is not prejudiced as a result of his employer failing to comply correctly with the Act.

It is normal to extend this policy insofar as the construction industry is concerned, to cover labour masters and/or their employees and/or self-employed persons and/or operatives hired in under a contract, to be deemed to be employees for the purpose of the insurance. It should be emphasised that this is done to protect the employer and does not, for example, protect the labour master in the event of a claim against him from one of his labour gang.

Public liability insurance is in respect of the employer's legal liability for death, injury, illness or disease to any person, other than an employee or anybody deemed to be an employee, which we have dealt with under the previous heading, and his liability for loss or damage to any property. Again, the construction industry does not have a particularly good record in respect of claims for death or injury or loss or damage as referred to herein. Consequently, the premium tends to be costly.

The construction industry as a consequence is not one of the insurance companies favourite types of business. Not all insurance companies will undertake this cover for members of the construction industry, and of those that do many provide policy workings with exclusions and limitations in the cover which make the policy unsuitable. Certainly, in those policies there are many areas where a contractor could find that unwittingly he is uninsured in the event of a particular set of circumstances occurring. It is not possible even to summarise in this section the restrictions as they are too numerous, and differ from insurer to insurer.

Some of the major areas of difficulty, however, are:

– an exclusion of liability assumed under contract or agreement. This is not acceptable as even the standard form of building contract or any of the sub-contracts which go with it impose some form of liability assumed under contract. Other such liabilities are imposed by various other agreements outside the main contract whether these be with local authorities, adjoining property owners, plant hirers, etc
– liability for loss or damage to property supplied by the insured. Here the true intention is to exclude the cost of putting right defective workmanship which no Public Liability insurance policy will cover. However, many of the policy wordings go far beyond the intention of exclusion of defective work and exclude the cost of putting right any sound work provided by the contractor which was damaged as a result of some defect elsewhere in that contractor's work. For example in some wordings, if the contractor was to provide two identical tower blocks one of which was built perfectly, but the other was defective in workmanship, materials or design, and as a result collapsed and in so doing collapsed on to the sound building, insurers would not pay anything towards the cost of reconstruction of either building on the basis that the loss had arisen out of the provision of the defective work. This leaves the contractor with an extremely wide gap in cover.

No insurer will pay the cost of putting right anything which is in itself defective, but cover is obtainable which merely excludes the cost of putting right the defective work.

It is normal for most building contracts to insist that the Public Liability risk is insured by the contractor for a minimum amount of insurance cover. It should be recognised, however, that the limit imposed in the contract is the minimum requirement and does not in any way restrict the amount which his employer may claim against him.

It is extremely difficult to obtain insurance providing adequate cover. This is really a matter to be discussed with the contractor's professional insurance advisers.

3.6 Professional indemnity

This policy is intended to cover the contractor against his legal liability arising out of a defect in design or specification carried out by him or for which he has accepted responsibility under the terms of a contract. Design and construct package deals are becoming more popular in the construction industry and the contractor either takes on the responsibility of doing the design in-house or alternatively whilst contracting on a design-and-build contract form and thereby accepting responsibility for the design, he sub-contracts the design work to other professionals.

The number of insurers providing this sort of cover is extremely small, which reflects the view of the insurance market in general of the risk involved. Many people are nowadays ready to challenge their professional advisers, and this appears to be particularly so where that professional advice is provided as part of a package to design-and-build. As a consequence the premiums are extremely costly and the contractor is always required to carry a fairly substantial self-insured amount, before the insurance policy will apply. The indemnity limit is more often than not provided on the basis of an aggregate figure during the course of any one period of insurance (as distinct from a limit on any one claim without any limit on the number of claims during the course of the period of insurance which is more normal with Employer's Liability and Public Liability insurance). Under Professional Indemnity insurance, therefore, the indemnity limit provided by the policy is reduced by the amount of any successful claim made against that policy.

To obtain cover it is necessary to complete a long and complicated proposal form designed to extract detailed information about the type of work carried out by the company, the type of design undertaken, the amount of design work undertaken, the qualifications of the persons doing the design, the previous design claims history, etc. It is extremely important that this form is completed accurately as in the event of any incorrect information in the form, insurers may, and do, endeavour to decline to pay a claim on the basis that the actual risk they were writing was not the same as presented to them in the proposal form.

More often than not a design-and-build contract will put the design responsibility firmly on to the contractor, and in addition may insist that responsibility is insured.

3.7 Motor insurance

is something that most people know something about, and is not dealt with in depth in this section. The only area of Motor Insurance which is covered is where there are special requirements in respect of the construction industry.

One of many difficulties with the construction industry is that it has in general a lot of plant other than normal road-going vehicles. It requires careful thought with regard to the arrangement of insurance as there are various ways of doing this. Cover may include loss or damage to the plant itself under the Motor policy in which case the contractor will probably be paying too high a premium. Alternatively cover can be arranged for the Motor policy to cover only legal liability with the accidental damage cover coming under the contractor's policy covering his contract works. The two policies need careful dovetailing as otherwise there is likely to be overlapping cover or, far more serious, a gap in cover between the two policies. It may be for example that the policy on the contract works will not cover accidental damage to plant whilst it is travelling under its own power on a public highway. In this case some accidental damage cover is still required under the Motor policy. Motor insurance is a complicated subject which requires careful arrangement.

3.8 Contract guarantee bonds

Technically, contract guarantee bonds have nothing to do with insurance. The only common link is that some insurers undertake to write contract guarantee or other bonds.

A contract guarantee bond is issued where the employer, generally although not always a local authority, requires as part of the contract conditions that the contractor provides a bond which the employer can invoke in the event that the contractor fails to perform his contract, subject to certain restrictions. Generally speaking in this country this would apply mainly in the event that the contractor went into liquidation. In this event the employer would call upon the surety, the company underwriting the bond (whether this be an insurance company or a bank) to pay up to the maximum stated percentage of the contract value in respect of the additional costs which the employer has incurred as a result of having to complete the contract through other contractors.

This is not insurance, and does not come within normal insurance law. Not all contractors would automatically qualify for a bond. By agreeing to issue a bond the surety is in fact saying that it believes that the contractor is financially sound and has all the resources to complete the contract in question. If the contractor fails to do that the surety company may well have to pay. Consequently it can be seen that the financial status of the contracting company is an all-important factor in deciding whether a bond is available to that contractor and consequently the surety will look very carefully at the company's accounts, its record and whether it has completed jobs of a similar nature in the past.

The company providing the bond will need to decide whether it believes that the contracting company has the ability to complete the work. If it does not believe that, the chances are that no amount of premium could buy a bond for that particular contract in those circumstances.

3.9 Premium

Underwriters and insurance companies are, like everybody else, in business to make a profit. One of the principles of insurance is that it is the premiums of the many which contribute to the losses of the few. However, with certain companies, particularly in the construction industry this principle becomes a little difficult to apply in that it is by no means the 'losses of the few' that we are dealing with, but very often a considerable number of losses under the same policy and the same year. Consequently, the insurer is endeavouring to assess in advance, from past experience, how much he feels he is likely to pay out in claims in the forthcoming 12 months, to which he must then add something for the cost of handling those claims, and profit, in order to arrive at an equitable premium.

The premium for Employer's Liability and Public Liability insurance is more often than not based at a rate on the contractor's wageroll. The rate will depend upon the individual contractor's record of accidents, and the claims paid by the insurance market in respect of those accidents. The construction industry in general does not have a good accident record, and consequently insurance rates are relatively higher in the construction industry than in many others.

Insurance against loss or damage to the works, etc, is more often than not calculated as a rate on the contract value or the value at risk in the event of insurance being arranged in respect of a single project or as a rate on the contractor's turnover if the policy is arranged on an annual renewable basis covering any work carried out during the course of the 12 months period. Rates vary considerably depending upon the claims experience of the individual contractor, and the extent of the cover actually being given by the policy, eg, whether the contractor's own plant is included or excluded, whether full cover is being given to sub-contractors or whether sub-contractors are responsible for their own works, etc.

The calculation of premium for Professional Indemnity insurance is extremely difficult to explain, but is based largely upon the amount of design responsibility actually undertaken by the contractor and whether the contractor has actually carried out that design work himself or has sub-contracted the work to other professionals against whom a separate claim could be brought in the event that design failure was caused by their negligence. The claims experience for the individual contractor is, again, a factor which will be taken into account.

The premium for contract guarantee bonds is rather different. As has been said above, if the company being asked to underwrite the bond does not feel the contractor has the financial or technical ability to complete the particular job which is the subject of the request for a bond, it is highly unlikely that any premium will buy such a bond. It is obvious, therefore, that the premium rate charged depends upon the underwriter's judgement of the financial and technical viability of the contractor to do the job, and is normally quoted as a rate on either the bond value or the contract value.

3.10 Self-insurance

Many insurance policies include something referred to as an excess, self-insured or deductible amount. All of these titles refer to the insured paying for the first portion of any loss which occurs. For example, under the policy covering loss or damage to the works, it would not be unusual to see an excess of £500 for each and every claim meaning that insurers would only pay for that proportion of the loss or damage which exceeded £500 arising out of any one incident.

It is possible to carry a higher excess on a voluntary basis in order to keep the premium rate to a minimum.

Bearing in mind that insurers are in business to make a profit there is little point in insuring against lots of small incidents because insurers will merely charge a premium to recoup these amounts plus a percentage to pay for the claims handling and profit. It is, therefore, often preferable to accept an excess higher than that actually imposed by insurers, which can be financially advantageous to the contractor opting to self-insure (other than for Employer's Liability insurance or Motor insurance which are compulsory by law). In certain circumstances this can be advantageous, but it must always be borne in mind that a catastrophe or series of catastrophes might not only deplete any insurance fund which had been set up to pay for these losses but could put the contractor into financial difficulty. Because the construction industry has a high hazard rating it is not normal for contractors to totally self-insure. Indeed, due to the terms of contracts they enter into it is often unacceptable to the principal.

3.11 Claims

So far we have considered only types of insurance and placing it but the whole point of insurance is to be able to recover in the event of some misfortune occurring causing financial loss.

There are certain factors which have to be considered when making an insurance claim. Firstly, the insurers by the terms of the policy will require notification of any loss or accident which may turn into a loss as soon as this is discovered by the insured, and will expect to be allowed to handle this claim on behalf of the insured. It is important, therefore, to advise each incident immediately it occurs.

Having reported the incident insurers will normally guide the insured as to what action they would wish them to take, and what information they will require. This will vary from policy to policy and indeed from incident to incident under some policies. Very often, but not always, a claim form will need to be completed, but these do not necessarily always bring out all the information which is necessary. Certainly therefore, on major losses insurers would like to see a written statement of what has happened, and how it occurred to the best of the insured's knowledge. It is important that this is a true and accurate statement and that no information is hidden as this may well inhibit the handling of the claim, and could in fact prejudice the insured's right to recover under the terms of the policy, if the insurer feels they have been prejudiced by wrong information supplied deliberately.

It is important to keep records of payments made in respect of restoration of damaged work, and similarly in the event of injury to persons to keep any plant, machinery or equipment which may have caused the injury as this may be needed as evidence by the insurers or even by the Courts in the event that such claim is actually brought to Court.

In larger claims insurers may well employ loss adjusters to seek out the circumstances and the financial details of any loss. In some cases this may be undertaken by solicitors. It is obviously important that the insured provides all the assistance to these parties who are acting as agents of the insurer.

3.12 Exercises

In the following exercises, it is assumed that a JCT 1980 contract with 1986 amendments is applicable, with nominated sub-contractors under NSC/4.

Consider the following:

1. Under the contract, whose responsibility is it to insure against a loss caused by fire to the works where the loss is sustained by:
   (a) the main contractor?
   (b) a nominated sub-contractor?
   (c) a domestic sub-contractor?

2. Whose responsibility to insure for accidental damage to the works of:
   (a) the main contractor?
   (b) a nominated sub-contractor?
   (c) a domestic sub-contractor?

3. What perils does the main contractor need to insure against, and under which policy if:
   (a) clause 22A is invoked?
   (b) clause 22B is invoked?
   (c) clause 22C is invoked?

4. What are the main policies required to be taken out in respect of insuring liabilities?

5. If the sub-contractor's works are damaged by a flood during the course of construction, who would be responsible for the cost of restoring the work, and under whose insurance policy and which type of policy would the claim be made?

6. A hoist collapses due to incorrect installation injuring employees of the main contractor and sub-contractors. Who would claim against whom and what insurance policies are likely to be involved?

7. Due to an error in design, part of the works collapses. Whose responsibility would this be, and what insurance may be available to pay?

8. Prior to practical completion, the works are taken over and fire damage occurs to the part taken over. Whose responsibility is it and whose insurance would pay?

9. *Discuss* the contractor's liability to insure under JCT 80.

10. A nominated sub-contractor carrying out piling work on an urban site experiences difficulties with the driving operation. In order to maintain the required rate of progress, it is found necessary to increase the size of hammer from that originally agreed with the architect. A short time after the change, fractures appear in the brick gable wall of an adjacent building.
    Discuss:
    (a) the obligation in respect of such damage under JCT 80
    (b) the insurance position.
    (Clause numbers are to be quoted where appropriate).

3.13 Sources and further reading

JCT forms of contract insurance clauses were revised significantly in 1987 so references to those contracts in the following should be read with those revisions in mind.

MANSON K, Insurance for the small builder, *Building Trades Journal* 1986, 192, Nov and Dec 27, p 20

MANSON K, Why you might be liable, *Chartered Builder*, November 1990

STUBBS D P, Public Liability insurance and its complexities. *Building, Technology & Management* April 1985, pp 31–32

SPEAIGHT A, and STONE G, *AJ Legal Handbook* Architectural Press, 1983, 3rd edition

BEC Contractor's insurance compendium (including Supplement) 1978

Section 4
RESOLUTION OF DISPUTES

This section is concerned with:
- alternative approaches to the resolution of disputes
- arbitration as the principal method; procedures, the award, fees, advantages and disadvantages.

4.1 Introduction

When the parties to a contract find they have a difference of opinion or a dispute arises there are two primary courses open to them under English law both of which involve submitting the facts of the dispute to a third party for him to settle.

The first course, litigation, involves a third party who is trained and qualified in the law; a judge appointed by the courts.

The second course, arbitration, involves a third party who is appointed because he is an expert in the matters in dispute. He may have some legal training or even hold a legal qualification but he is appointed because of his expertise in the matters in dispute rather than for his legal knowledge.

Definitions of litigation and arbitration are given in section 4.2.

Under what circumstances is one of the above courses used in preference to the other?

If the parties in dispute, the disputants, have a contract which does not include expressed conditions regarding the method of settling a dispute, should one arise, they will probably apply to the courts. This is the direction in which solicitors usually look. Indeed, the case may well be heard in the County or High Court, depending on the sum involved.

If, however, the matter in dispute is primarily of a technical nature despite the fact that points of law may be involved the High Court will probably refer it to the official referees.

An official referee is a judge but his court is more adaptable to commercial procedures than the County or High Courts. The procedures followed in his court are similar to some of those described below. Whilst the official referee is unlikely to be an expert in the matter in dispute if it is of a technical nature, he has frequently devoted considerable time to 'commodity' disputes and many building disputes are resolved in the court of the official referees.

A significant number of disputes, particularly those concerned with consultants' fees and builders' accounts on smaller projects are, however, referred by the courts for settlement by arbitration.

A problem facing disputants is that English courts have, for many years, taken months or even years before they have been able to deal with many cases so that building contracts which do not have conditions of contract requiring disputes to be referred to arbitration for settlement may experience significant delays before being heard. The majority of building projects of any magnitude do, however, have arbitration clauses so delays by the courts in this respect do not arise.

The JCT, ICE and most if not all contracts for public works contain conditions which require arbitration as the method.

Arbitration is, then, a much used procedure for building disputes and it is that course of action with which this section is principally concerned. There are, however, a number of approaches which may be used when a difference of opinion occurs but when the parties do not yet have a dispute on their hands and it is with these that section 4.2 is mainly concerned.

4.2 Disputes and differences – alternative approaches for resolution

As far as the parties to building contracts are concerned, disputes are best avoided or resolved amicably, between the parties.

Norman Royce, perhaps the most experienced and best known arbitrator in construction disputes, a person one would expect to have a vested interest in promoting settlement of disputes by arbitration recommends that the parties make every effort to resolve their differences before they develop into disputes. In a paper concerned with alternative approaches to dispute resolution, before embarking on arbitration or litigation, he suggests:

'... the most desirable way of resolving any dispute is that it should enjoy the confidence of the parties as a method likely to arrive at a just answer... The most desirable way of resolving any dispute is for the parties themselves to reach a mutually acceptable compromise. This is likely to be quicker and cheaper; no third party may be involved or even informed of the dispute; and future business relations can be maintained.'[1]

There are, however, occasions when differences of opinion and disagreements pass beyond the point where the parties to a contract feel that they are able to reach a mutually acceptable settlement and they decide that reference should be made to an independent third party.

What are the courses open to them?

There are several options available to the parties to a contract in the event of a dispute or difference occurring. To some extent the process adopted depends on the extent to which the attitudes of the parties, the *disputants*, have hardened and their commitment to resolving the dispute without recourse to law.

The following approaches are available to them as a difference of opinion hardens and becomes a dispute.

Conciliation 'Conciliate. Gain (esteem, goodwill); pacify; win over (to one's side, etc); reconcile.' (OED)

Conciliation involves 'use of conciliating measures' with 'Court of Conciliation (offering parties a voluntary settlement)' (OED) or the introduction of a third party who has not been involved between the parties to a dispute in order to clarify the issues and bring about resolution of their differences.

Various bodies have conciliation rules aimed at bringing together the parties to business disputes. ACAS is a statutory service charged with performing a similar role with regard to industrial disputes.

The Society of Construction Arbitrators is considering progressing with conciliation and CIArb has guidelines in draft.

Mediation 'Mediate. Form connecting link between; intervene (between two persons) for purpose of reconciling them.' (OED) Conciliation and mediation have essentially the same meaning. Mediation is the term more frequently used in the USA where methods for 'alternative dispute resolution' have been developed more rapidly than in Britain.

Mini-trial Mini-trial is a misleading term in that it is not a trial but a structured non-binding settlement which has gained popularity in the USA in recent years. Various options exist but essentially the mini-trial involves the lawyers for each party making a brief presentation of the case to executive members of the firm who have authority to settle the dispute. The aim is to provide them with an opportunity to hear the other side's case and make an assessment of risk. Having observed the presentation the executives meet together to discuss settlement with the 'business' facts fresh in their minds.

A third party may be involved to supervise the presentations and, if requested, give an opinion on the possible outcome of the 'case' should it go to trial by litigation or arbitration. The mini-trial is usually concluded within one day.

Advantages for the disputants are privacy, informality, speed and economy. It is doubtful if the mini-trial would prove popular with the legal profession.

Adjudication 'Adjudicate (of a judge or court). Decide upon (claim etc.).' (OED)

Adjudication involves the introduction of a third party who considers evidence put to him by the disputants and makes an award. The disputants are not bound to accept the adjudicator's award. There is provision for adjudication in many JCT forms of subcontract and in some other standard forms of contract between client and contractor, but as a process, adjudication does not have statutory backing in the same way that the following courses do.

The BPF system, discussed in section 5.2, also provides adjudication as a means of resolving disputes, to paraphrase the BPF manual, quickly rather than letting them become prolonged and harmful to cooperation and the project. BPF suggests that the adjudicator should be conversant with building costs and methods and with estimating and administration matters. He should act as a conciliator wherever possible. Unlike the arbitrator, the adjudicator is appointed at the beginning of the contract and he is expected to be available at short notice.

If either party is not satisfied with the adjudicator's decision, the dispute may be referred to arbitration. The adjudicator's decision forms part of the evidence to the arbitrator.

Ligitation is 'the action or process of carrying on a suit in law or equity'. *In litigation* means 'in process of investigation before a court of law' (OED).

Arbitration To arbitrate is 'to examine, give judgement' (OED). The powers of the arbitrator and the procedures to be adopted are defined in the Arbitration Acts of 1950, 1975 and 1979. Parties entering arbitration are bound to accept the arbitration award as final and binding upon them. The award is as enforceable as a judgement of the court.

Which approach to use? The above options are listed in increasing order of formality and legal enforceability.

Neither conciliation nor mediation involves the conciliator or mediator making an award. The aim is to bring the parties together before their respective attitudes have hardened. Before, in effect, they have become *disputants*. Goodwill is an important ingredient of successful conciliation and/or mediation.

Most of the bodies publishing conciliation rules require the parties to conciliate before resorting to arbitration. Indeed, ACAS must not refer to arbitration unless conciliation and other associated procedures have failed.

Royce, too, believes it is only when conciliation fails that arbitration should proceed but he suggests that, if the dispute is at all complicated, the following are the prerequisites for successful conciliation between the parties.[2]

(i) They must have a genuine desire for a fair settlement with the other(s) involved in the difference and must not be seeking a victory which is based on legal niceties. They must realise that any settlement reached by conciliation is not legally enforceable and that its honouring depends solely on the good faith of the parties.

(ii) They must be mutually seeking a method of settlement which is capable of avoiding the additional costs, which are unavoidable in references to the courts or to arbitration, due to the duplication of legal processing, advocacy and the experts' opinions in such references.

(iii) They must be seeking a settlement process which is quicker than is practicable by reference to the courts or to arbitration.

(iv) They must appreciate that an

attempt to reach a fair settlement by conciliation does not prevent resort to the courts or to arbitration if the conciliation fails, and that the assembly of the particulars needed for conciliation will, in all probability, reduce the time involved in any later reference to the courts or to arbitration.

(v) They must appreciate that the assembly of the particulars required for conciliation by a skilled third party will often of itself induce a settlement – thus reducing the abortive expense which occurs in the large number of instances where court cases and arbitration references are settled before the dispute reaches a hearing.

(vi) They must be willing to disclose to the conciliator in confidence guaranteed by the conciliator, all particulars relating to the matters in difference which they would disclose to their own representatives in court or arbitration proceedings.

(vii) They must be willing to deal personally with the conciliator without legal representatives and only with technical representatives if the conciliator's considered consent to this is appropriate, and the other party consents.

(viii) They must understand that the conciliator cannot after the conciliation proceedings have terminated, be appointed as an arbitrator or act as advocate.

(ix) They must understand that either party to the difference can terminate the conciliation at any time; and that termination by either party terminates the conciliation proceedings.

Adjudication, as the definition above indicates, can be a more formal and legally binding process that conciliation and mediation but even adjudication does not have the weight of case law and statute behind it in the same way as litigation and arbitration.

As stated in section 4.1, litigation is the normal course of civil action for disputants but many contracts for construction works provide for the settlement of disputes or differences by arbitration and it is with arbitration that this section is concerned.

The need to submit a dispute to litigation or arbitration should be regarded as an indication of failure by the parties who should seek to resolve differences they may have by any of the various approaches outlined above before, ultimately, resorting to arbitration which is all too often an expensive and lengthy process.

4.3 Arbitration – the legal framework

Figure 4.1 illustrates the structure and relationship of English courts relating to civil disputes.

Arbitration in England and Wales is controlled by the Arbitration Acts of 1950, 1975 and 1979. The 1979 Act is referred to as 'an Act to amend the law relating to arbitrations and for purposes connected therein' and Section 1 (i) of that Act refers to the 1950 Act as 'the principal Act'. The 1975 Act appears on the face of it to refer to the enforcement of foreign arbitration awards but it can be relevant to domestic arbitrations where there is an application to stay court proceedings in favour of arbitration and one of the parties is a foreign national or a foreign company, even though in all other respects the contract is domestic.

The scope of the 1950 Act may be ascertained from a study of the 'arrangement of sections, Part 1' which is set out in full, below:

General Provisions as to Arbitration
Effect of Arbitration Agreements, etc

Section
1   Authority of arbitrators and umpires to be irrevocable
2   Death of party
3   Bankruptcy
4   Staying court proceedings where there is submission to arbitration
5   Reference to interpleader issues to arbitration.

ARBITRATION – THE LEGAL FRAMEWORK

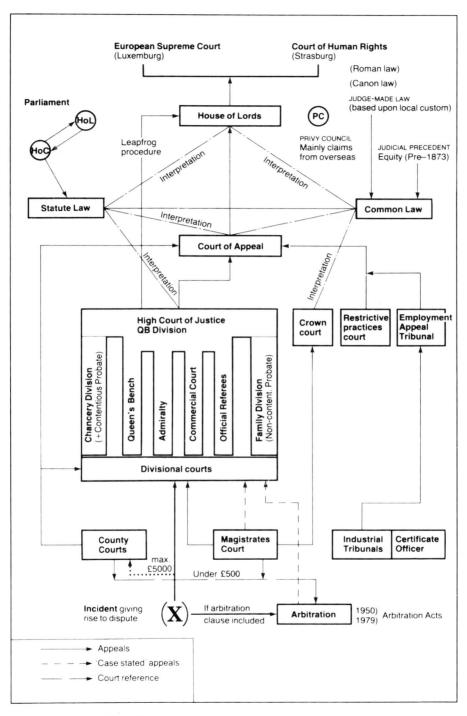

4.1 *Structure and relationship of English courts relating to civil disputes*
BADEN-HELLARD, R, 1988

Arbitrators and umpires
6 When reference is to a single arbitrator
7 Power of parties in certain cases to supply vacancy
8 Umpires
9 Agreements for reference to three arbitrators
10 Power of court in certain cases to appoint an arbitrator or umpire
11 Reference to official referee.

Conduct of Proceedings, Witnesses, etc
12 Conduct of proceedings, witnesses, etc

Provisions as to awards
13 Time for making award
14 Interim awards
15 Specific performance
16 Awards to be final
17 Power to correct slips

Costs, Fees and Interest
18 Costs
19 Taxation of arbitrator's or umpire's fees
19A Power of arbitrator to award interest
20 Interest on awards

Special cases, Remission and setting aside of awards, etc
21 Statement of case
22 Power to remit award
23 Removal of arbitrator and setting aside of award
24 Power of court to give relief where arbitrator is not impartial or the dispute involves question of fraud
25 Power of court where arbitrator is removed or authority of arbitrator is revoked

Enforcement of award
26 Enforcement of award

Miscellaneous
27 Power of court to extend time for commencing arbitration proceedings
28 Terms as to costs, etc
29 Extension of s.496 of the Merchant Shipping Act, 1894
30 Crown to be bound
31 Application of Part I to statutory arbitrations
32 Meaning of 'arbitration agreement'
33 Operation of Part I
34 Extent of Part I.

Part II of the Act is concerned with 'Enforcement of certain Foreign Awards' and Part III with:

— the short title of the Act (Arbitration Act, 1950)
— repeal of the 1889, 1924 and 1934 Acts
— two schedules which contain agreements between the states comprising the League of Nations with regard to arbitration.

The titles of the sections indicate how comprehensive the Act is in its coverage. The effect of arbitration agreements, arbitrators and umpires, conduct of proceedings, provisions as to awards, costs and fees are all included.

One of the aims of the 1979 Act was to remove causes of complaints from foreign business men which might lead to a reduction in England's pre-eminence as an arbitration centre.

The scope of the 1979 Act may also be ascertained from a study of the arrangement of the sections:

Section
1 Judicial review of arbitration awards
2 Determination of preliminary point of law by court
3 Exclusion agreements affecting rights under sections 1 and 2
4 Exclusion agreements not to apply in certain cases
5 Interlocutory orders
6 Minor amendments relating to awards and appointment of arbitrators and umpires
7 Application and interpretation of certain provisions of Part I of principal Act
8 Short title, commencement, repeals and extent

Reference was made above to the 1979 Act amending the law relating to arbitrations.

The 1979 Act makes three significant changes from the 1950 Act.[3]

(a) *Special case procedure*
The first is a modification of the special case procedure. Before the 1979 Act any point of law arising during or out of an arbitration could be referred to the High Court for a decision of the

Court which he could then take into account when making his award. The procedure had advantages but it tended to be over-used with the result that a backlog of cases awaiting trial developed and parties experienced cash-flow problems while awaiting a decision.

The special case procedure was abolished in the 1979 Act and a right of appeal to the court was introduced with provision for a consultative case to be considered before an award was made. In certain circumstances the parties were given the right to exclude the Court's jurisdiction.

There is a considerable body of opinion that the special case procedure has not, as hoped, been very successful in reducing the number of cases submitted to the Court.

(b) *Reasoned award*
The second change made provision for the arbitrator to give 'reasoned awards'. Previously, arbitrators generally made their awards without stating the reasons for them.

The 1979 Act did not make reasoned awards obligatory but it gave the Court power to order an arbitrator to give his reasons where an appeal is pending. In most cases the arbitrator must be given notice that reasons will be required. Reasoned awards are discussed in section 4.8.

(c) *Exclusion of judicial review*
The principle adopted is that so far as might be possible and just an arbitral award should be final. If an award is not final 'the main point and purpose of commercial arbitration may become so diluted as to make the whole procedure of little value to men of business'. The new Act lays down clear rules for the timing and scope of 'exclusion agreements'. An exclusion agreement is one which excludes the parties' right to appeal on a point of law in the award.

Traditionally, English arbitrators have been reluctant to give reasoned awards which the Court had the power to set aside for an error of fact or law. The 1979 Act restricts the powers of the Court to set aside.

The effect of the Acts
The 1950, 1975 and 1979 Arbitration Acts provide the statutory framework for the arbitral process but contractual agreements may impose conditions which are outside the Acts. The JCT Arbitration Rules, 1988, discussed in section 4.4, provide an example of rules which supersede the Acts.

4.4 Arbitration as a standard condition of contract

The JCT Standard Forms of Building Contract contain conditions of contract which provide for arbitration as the basis for the settlement of disputes.

In addition to the conditions of contract, the majority of the forms of contract published by the JCT are subject to the JCT Arbitration Rules, 1988. These rules have contractual effect.

The rules are concerned with:

– the service of statements, documents and notices and the content of the statements
– the manner in which the arbitration is to be conducted (procedure without hearing, full procedure with hearing or short procedure with hearing)
– inspection by the arbitrator
– arbitrator's fees and expenses – costs
– payment to trustee – shareholder
– the award
– the powers of the arbitrator.

The Arbitration Rules were a response by JCT to criticisms from clients and the industry that arbitration was becoming too costly and an over-lengthy process. One of the most significant aims of the rules is to reduce costs and shorten the time taken by the traditional arbitral process. JCT 80 conditions of contract and the Arbitration

Rules have been used, below, to illustrate the process and procedures.

4.5 Four procedures in outline

Four procedures are discussed below. The first, which will be referred to as the 'traditional' procedure, has been used as a datum from which to consider the others.

The traditional process The steps in the traditional process may be summarised as follows:

1. two or more of the parties to the contract or sub-contract are in a dispute
2. the parties agree to submit their dispute to arbitration
3. an arbitrator is appointed
4. arbitrator calls preliminary meeting of the parties concerned to discuss arrangements and requirements regarding submissions from the parties. These are referred to in steps 6–10, below
5. the party initiating the dispute, the claimant, prepares points of claim for submission to the other party/ies, the respondent/s
6. respondent prepares points of defence and/or points of counterclaim for submission to claimant
7. in the event of the pleadings (collective name for the points of claim, defence and counterclaim) being insufficiently detailed either party makes a request for further and better particulars
8. either party may examine the property involved in the dispute
9. in the event of either party being reluctant to disclose the property, the arbitrator may make an order for discovery and/or inspection
10. when claimant, respondent and arbitrator are agreed as to the scope of the pleadings, arbitrator arranges the hearing at which the procedure is as shown in steps 11–20, below
11. claimant sets out his case on the claim and counterclaim, if any
12. claimant calls his witnesses and examines them
13. witnesses are cross-examined by respondent and/or questioned by arbitrator
14. respondent sets out his case
15. respondent calls his witnesses and examines them
16. witnesses are cross-examined by claimant and/or questioned by arbitrator
17. respondent sums up his case in his closing speech
18. claimant replies to respondent
19. arbitrator closes hearing
20. arbitrator may inspect the works
21. arbitrator make his decision
22. arbitrator serves his award
23. award is enforceable as judgement debt.

The above steps should be kept in mind when considering in outline the three procedures contained in the JCT Arbitration Rules and, again, when considering the procedures in more detail in section 4.6.

A more detailed sequence of events is shown in the flow chart contained in figure 4.2.

The procedures in the rules
The alternative procedures contained in the rules are:
– rule 5, procedure without hearing
– rule 6, full procedure with hearing
– rule 7, short procedure with hearing.

Which procedure to use?
To some extent the initiative rests with the claimant who within a reasonable time after the commencement of the arbitration and at least 7 days before a decision under rule 4.2.1 is required, to formulate his case in writing in sufficient detail to identify the matters in dispute and submit that written case to the respondent with a copy to the arbitrator and state that at the preliminary meeting he will request the respondent to agree that rule 7 shall apply to the conduct of the arbitration.

If, at the preliminary meeting the parties agree, the provisions of rule 7 will apply and the arbitrator must issue a direction to that effect.

FOUR PROCEDURES IN OUTLINE

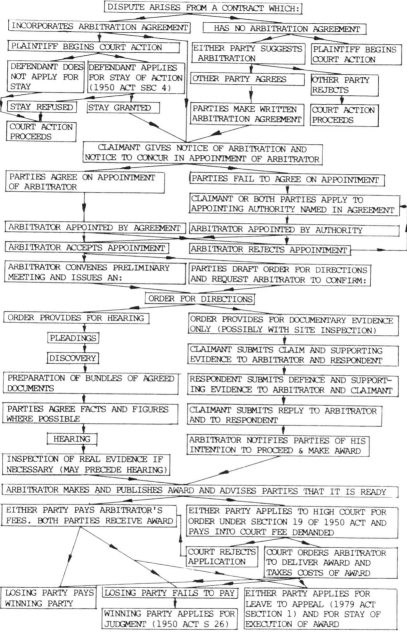

4.2 *Flow chart for arbitration*
   STEPHENSON, D A, 1987

If, at the preliminary meeting the parties do not agree the arbitrator must issue a direction as to whether rule 5 or rule 6 will apply (rule 4.2.2).

Rule 4.3.1 empowers the arbitrator to direct that rule 5 will apply, unless, having regard to any information supplied by and/or representation made by the parties, he directs that rule 6 will apply.

There is, then, a presumption towards a procedure *without hearing* unless the parties persuade the arbitrator that a hearing is preferable.

A direction given under rule 4.3.1 must be issued within 28 days of the Notification Date or, if a preliminary meeting has been held, not later than 7 days after the date of the preliminary meeting (rule 4.3.2). Whichever of the procedures is adopted the rules so far as relevant and applicable apply (rule 4.3.3).

Procedure without hearing (rule 5) The procedure without hearing runs to a tight programme of 14 days 'legs' each of which is conditional on the one before.

The initiative rests with the claimant who must serve a statement of case within 14 days of the rule 5 procedure becoming effective.

The respondent is required to serve his statement of defence and, if any, his statement of counterclaim within 14 days of the service of the claimant's statement of case.

The claimant has a further 14 days from service of a statement of counterclaim by the respondent to serve a statement of defence to the counterclaim.

The last leg is achieved if the claimant serves a statement of defence to the respondent's statement of counterclaim within the time or times allowed by the rules. The respondent may, within 14 days after such a service, serve a statement of reply to the defence.

As may be seen in figure 4.3, the procedure runs to 56 days if each 14 days period is used in full.

The parties serve their various statements and other documents on the arbitration and on each other.

The claimant provides his statement of case and any statement setting out a reply to the respondent's statement of defence and his statement of defence to any statement of counterclaim by the respondent.

The respondent provides his statement of defence, any statement of counterclaim and any statement setting out a reply to the claimant's statement of defence to any counterclaim.

Either or both parties also provide a list of documents which they consider necessary to support any part of the relevant statement and a copy of the documents identifying clearly in each document the parts which they regard as relevant to the case.

If either party fails to keep to the timetable rule 5.4 empowers the arbitrator to notify the parties that he proposes to proceed on the basis that the party will not be

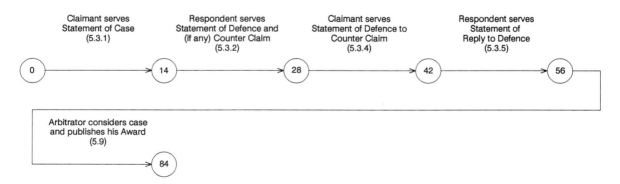

4.3 *Procedure without hearing – the timetable (the figures in the nodes represent days from commencement)*

serving the same unless within 7 days of the date of service of that notification the relevant statement is served. If within 7 days of the date of service of that notification the relevant statement is not received the arbitrator proceeds on the basis that the party will not be serving the same.

If a statement is served late it is of no effect unless the arbitrator is satisfied there was a good and proper reason for the lateness of service.

If the claimant does not serve his statement of case in time the arbitrator is obliged to make an award dismissing the claim and ordering the claimant to pay his fees and expenses and any costs hitherto incurred by the respondent (rule 5.5).

Either party may, within 5 days of service of documents on the arbitrator and the other party, serve written comments on the other party's submission to the arbitrator which the arbitrator must take into account when exercising his discretion.

If the arbitrator considers that he requires clarification by an interview with or further documentation from either of the parties he may require such clarification but in doing so he must take care that the other party is provided with the same clarification.

The arbitrator is required to publish his award within 28 days after receipt of the last of the statements and documents or after the expiry of the last of the times allowed by the rules for their service, whichever is the earlier.

Provided that he notifies the parties of his intention, the arbitrator may decide to publish his award later than the 28 days referred to above.

Full procedure with hearing (rule 6)
When the full procedure with hearing is adopted the claimant has 28 days to serve a statement of case. The respondent has 28 days after service of the claimant's statement of case to serve a statement of defence and a statement of any counterclaim.

The claimant may, within 14 days after such service, serve a statement of reply to any defence.

The claimant has 28 days after service of the respondent's statement of counterclaim to serve a statement of defence to the counterclaim.

The respondent has 14 days after service of any statement of defence by the claimant to serve a statement of reply to the defence.

The procedure with regard to the service by the parties of their various statements and other documents is essentially that required in rule 5, above and the arbitrator has essentially the same powers to proceed should either party fail to keep to the timetable.

After receipt of the last of any statements and documents or after the expiry of the last of the times allowed by the rules for their service, whichever is the earlier, after consultation with the parties, the arbitrator is obliged to notify the parties in writing of the date(s) when and the place where the oral hearing will be held. He is obliged to immediately notify the parties in writing of any change in dates or place (rule 6.8).

His award must be published within 28 days of the close of the hearing and he has the same power to delay publication of his award as he has with the procedure without hearing.

Short procedure with hearing (rule 7)
When the short procedure with hearing is adopted each party bears his own costs unless for special reasons the arbitrator at his discretion otherwise directs.

The hearing is held within 21 days of the rule 7 procedure becoming effective at the time, date and place directed by the arbitrator.

Not later than 7 days before the hearing, documents necessary to support the oral submissions (with relevant parts of them identified) are served on the arbitrator and other party.

The arbitrator directs the procedures to be adopted at the hearing and documents to be made available. No evidence except these documents may be adduced at the hearing except as the arbitrator may direct or allows.

At the end of the hearing the arbitrator is obliged either:

- to make his award which, if made orally, must be confirmed in writing or
- publish his award within 7 days of the hearing.

He has the same power to delay publication of his award as he has with the procedures described above.

The JCT Arbitration Rules provide for a procedure *without a hearing* in which event steps 10 to 20 would not apply. The rules have been drafted by the JCT as a basis for good practice and the three alternative procedures are discussed below.

It may be seen from above reviews of the three JCT arbitral procedures that the full procedure with hearing, as rule 6, most closely resembles the traditional procedure.

Having reviewed the procedures in context it is appropriate to look more closely at the individual steps in the arbitral process.

4.6 Procedures leading to the hearing – in more detail

The following items refer, in particular, to the traditional procedures. The appropriate clauses in JCT 80 and the JCT Arbitration Rules have been used as a basis.

The dispute, agreement to submit to arbitration and appointment of arbitrator (steps 1 to 3) The JCT Arbitration Rules (referred to below as 'the Rules') contain definitions which assist with an appreciation of the procedures.

Definitions – parties, time and notification date Rule 2.1 defines the *claimant* as the party who has required a dispute to be referred to arbitration and the other party as the *respondent*. Where the arbitrator has been appointed on a joint arbitration it is he who decides who will be claimant and who will be respondent.

The term *disputants* is frequently used to refer, collectively, to the claimant and the respondent (and has been used above) but the rules refer to them as the *parties* and that term will be used below.

An aim of the rules is to reduce the time taken to reach settlement of a dispute. Rule 2.3 provides that 'days' means calendar days but in computing any period of days, all public holidays, the four days following Easter Monday, December 24, 27, 28, 29, 30 and 31 shall be excluded.

The date when the arbitrator notified the parties of his acceptance of the appointment to proceed with the reference is known as the *notification date* (rule 2.5).

Rule 2.7 states that no time required by the rules, or by any direction of the arbitrator, may be extended by agreement of the parties without express written concurrence of the arbitrator.

Rule 2.7 gives the arbitrator teeth and should enable him to enforce a timetable on the parties and so achieve the aim of the rules.

Having in mind the importance of time and the method of computing time set out in rule 2 it would appear prudent for the arbitrator to set out a calendar stating dates by which each of the procedures is to be completed.

The arbitration agreement The agreement to submit disputes or differences to arbitration is contained in Article 5 of JCT 80 which reads:

'If any dispute or difference as to the construction of this contract or any matter or thing of whatsoever nature arising thereunder or in connection therewith shall arise between the Employer or the Architect on his behalf and the Contractor either during the progress or after the completion or abandonment of the Works, except under clause 31 (statutory tax deduction scheme) to the extent provided on clause 31.9 or under clause 3 of the VAT Agreement, it shall be and is hereby referred to arbitration in accordance with clause 41'.

Article 5 is clear in its meaning. For purposes of this study the reference to clause 31.9 may be ignored.

Clause 41, referred to above, sets out the procedures to be followed. It is contained in Part 4 of JCT 80 which is titled 'Settlement of disputes – arbitration'.

Reference to arbitration Clause 41.1 reads:

'When the Employer or the Contractor require a dispute or differences as referred to in Article 5 including:
- any matter or thing left by this Contract to the discretion of the Architect, or
- the withholding by the Architect of any certificate to which the Contractor may claim to be entitled, or
- the adjustment of the Contract Sum under clause 30.6.2, or
- the rights and liabilities of the parties under clauses 27, 28, 32 or 33, or
- unreasonable withholding of consent or agreement by the Employer or the Architect on his behalf or by the Contractor to be referred to arbitration then either the Employer or the Contractor shall give written notice to the other to such effect and such dispute or difference shall be referred to the arbitration and final decision of a person to be agreed between the parties as the Arbitrator, or, upon failure so to agree within 14 days after the date of the aforesaid written notice, of a person to be appointed as the Arbitrator on the request of either the Employer or the Contractor by the person named in the appendix.'

The clauses referred to in clause 41.1 are concerned with:

- clause 30.6.2, adjustments of the contract sum arising from prime cost sums, variations, nominated sub-contractors and nominated suppliers
- clause 27 and 28, determination of the contract by either the employer or the contractor
- clause 32, outbreak of hostilities involving the United Kingdom
- clause 33, war damage

Clearly, the scope of the matters which may be referred to arbitration is wide.

Opening the reference At what stage in the contract may the disputants *open the reference* (commence proceedings)?

JCT 80 clause 41.3 states:

41.3 Such reference, except
 .1 on article 3 or article 4; or
 .2 on the questions whether or not the issue of an instruction is empowered by the Conditions; or whether or not a certificate has been improperly withheld; or whether a certificate is not in accordance with the Conditions; or whether a determination under clause 22C.4.3.1 will be just and equitable; or
 .3 on any dispute or difference under clause 4.1 in regard to a reasonable objection by the Contractor, under clause 18.1 or clause 23.3.2 in regard to withholding of consent by the Contractor and clauses 25, 32 and 33,

shall not be opened until after Practical Completion or alleged Practical Completion of the Works or termination or alleged termination of the Contractor's employment under this Contract or abandonment of the Works, unless with the written consent of the Employer or the Architect on his behalf and the Contractor.

The article and clause numbers referred to above are concerned with:

- articles 3 and 4; appointment of a new architect or quantity surveyor in the event of the death of those named in the articles
- clause 22C 4.3.1; insurance of existing structures, determination of the employment of the contractor
- clause 4.1; the power of the architect to issue an instruction
- clause 18.1; partial possession by the employer
- clause 23.3.2; occupation of the site and the works, or part thereof, by the employer before the date of issue of the Certificate of Practical Completion
- clause 25; extension of time
- clause 32; outbreak of hostilities
- clause 33; war damage.

It will be apparent from the above list that whilst a commendable number of matters may be referred before Practical Completion, matters such as claims by the contractor due to him suffering loss and expense because the regular progress of the works has been disturbed must wait.

Appointment of arbitrator – by agreement or nomination? Having considered *when* an arbitrator may be appointed it is necessary to consider *how* he is appointed.

The closing paragraph of clause 41.1 assumes, first, that there will be agreement between the disputants as to the person who will act as arbitrator. Should they fail to agree on a person either party may request the president or a vice-president of the RIBA, RICS or CIArb to appoint a person.

The rules refer to the date of notification by the arbitrator to the parties of his acceptance of the appointment to proceed with the reference as the 'Notification Date'.

The arbitrator's qualifications, expertise and skills What qualifications, expertise and skills should the disputants seek when appointing an arbitrator?

A reason for using arbitration rather than litigation is that the parties may choose the person to whom they wish to submit their dispute. An arrangement which is not available to litigants. They should, therefore, take care to select a person with skills in the matters under dispute. If the matters under dispute cover different subjects they may be submitted to different arbitrators. If, for example, one matter is the failure of structural steelwork a structural engineer may have the appropriate expertise. If, as is frequently the case, the matter under dispute is the valuation of loss and expense suffered by the contractor due to disruption of the regular progress of the works, (a JCT 80 clause 26 matter), a quantity surveyor may be the appropriate person. There is no reason why the person selected should hold a professional qualification. It is expertise in the matter under dispute which is of primary importance.

The disputants have, then, the matter of selection in their own hands but should they fail to reach agreement they may receive some reassurance from knowing that the presidents referred to above keep lists of their members with specialist expertise who are prepared to act as arbitrators from which they, the presidents, may make their nomination. The lists contain details of the areas of specialist expertise of those named.

Section 12 of the 1950 Act titled 'Conduct of proceedings, witnesses, etc' states that the arbitrator has power to 'administer oaths to, or take the affirmations of, the parties to and witnesses on a reference under the agreement' (section 12(3)).

In addition, section 12(6) states:

'The High Court shall have, for the purpose of and in relation to a reference, the same power of making orders in respect of:

(a) security for costs
(b) discovery of documents and interrogatories
(c) the giving of evidence by affidavit
(d) examination on oath of any witness before an officer of the High Court or any other person, and the issue of a commission or request for the examination of a witness out of the jurisdiction
(e) the preservation, interim custody or sale of any goods which are the subject matter of the reference
(f) securing the amount in dispute in the reference
(g) the detention, preservation or inspection of any property or thing which is the subject of the reference or as to which any question may arise therein, and authorising for any of the purposes aforesaid any persons to enter upon or into any land or building in the possession of any party to the reference, or authorising any samples to be taken or any observation to be made or experiment to be tried which may be necessary or expedient for the purpose of obtaining full information or evidence; and
(h) interim injunctions or the appointment of a receiver

as it has for the purpose of and in relation to an action or matter in the High Court:

provided that nothing in this subsection shall be taken to prejudice any power which may be vested in an arbitrator or umpire of making orders with respect to any of the matters aforesaid.'

The 1979 Act did not alter the earlier Act insofar as section 12 of the 1950 Act is concerned.

The extract from the Act quoted above refers to action to be taken by the High Court or a judge because the Act may be used in connection with disputes arising on contracts which have no conditions for submission of disputes to arbitration.

More specific for building contracts are JCT 80 clauses 41.4 and 41.5 which read:

41.4 Subject to the provisions of clauses 4.2, 30.9, 38.4.3, 39.5.3 and 40.5 the Arbitrator shall, without prejudice to the generality of his powers, have power to rectify the contract so that it accurately reflects the true agreement made by the Employer and the Contractor, to direct such measurements and/or valuations as may in his opinion be desirable in order to determine the rights of the parties and to ascertain and award any sum which ought to have been the subject of or included in any certificate and to open up, review and revise any certificate, opinion, decision (except, where clause 8.4 is relevant, a decision of the Architect to issue instructions pursuant to clause 8.4.1), requirement or notice and to determine all matters in dispute which shall be submitted to him in the same manner as if no such certificate, opinion, decision, requirement or notice had been given.

41.5 Subject to clause 41.6 the award of such Arbitrator shall be final and binding on the parties.

The clauses referred to in the above are concerned with:

4.2	appointment of arbitrator to determine if the architect is empowered to issue an instruction, if the contractor so requests
30.9	the effect of the Final Certificate
38.4 .3	disagreement in connection with contribution, levy and tax fluctuations
39.5 .3	disagreement in connection with labour and materials cost and tax fluctuations
40.5	disagreement in connection with use of price adjustment formula.
8.4 & 8.5	disagreement in connection with materials, goods and workmanship to conform to description, and
41.6	agreement and consent by the parties pursuant to sections 1(3) (a) and 2(1) (b) of the 1979 Act to apply to the High Court to determine any question of law in the course of the reference and to appeal to the High Court on any question of law arising out of an award made in arbitration under the JCT 80 agreement. The parties agree that the High Court should have jurisdiction to determine any question of law.

A comprehensive statement of the powers of the arbitrator is contained in the Rules, rule 12 of which reads:

Powers of Arbitrator

12.1 In addition to any other powers conferred by law, the Arbitrator shall have the following powers:
  .1 after consultation with the parties to take legal or technical advice on any matter arising out of or in connection with the arbitration;
  .2 to give directions for protecting, storing, securing or disposing of property the subject of the dispute, at the expense of the parties or of either of them;
  .3 to order that the Claimant or Counter-Claimant give security for the costs of the arbitration or any part thereof, and/or for the fees and expenses of the Arbitrator, in such form and of such amount as the Arbitrator may determine;
  .4 to proceed in the absence of a party or his representative provided that reasonable notice of the Arbitrator's intention to proceed has been given to that party in accordance with the provisions of these Rules, including if there is to be a hearing, notice of the date and place thereof;
  .5 at his discretion to direct that the costs, if not agreed, shall be taxed by the Arbitrator;
  .6 to direct the giving of evidence by affidavit;
  .7 to order any party to produce to the Arbitrator, and to the other party for inspection, and to supply copies of any documents or classes of documents in the possession power or custody of the party which the Arbitrator determines to be relevant.

12.2 Subject to the Arbitration Acts 1950 to 1979 any noncompliance by the Arbitrator with these Rules, including those relating to time, shall not of itself affect the validity of an award.

12.3 If during the arbitration it appears to the Arbitrator to be necessary for the just and expeditious determination of the dispute that a Rule for the conduct of the arbitration other than that previously applicable shall apply, the Arbitrator, after considering any representations made by the parties, may so direct and shall give such further directions as he may deem appropriate.

Two further clauses in JCT 80 are concerned with the conduct of arbitration procedures and with cessation of his powers before making his final award. They are:

41.7 Whatever the nationality, residence or domicile of the Employer, the Contractor, any Sub-Contractor or supplier or the Arbitrator, and wherever the Works or any part thereof are situated, the law of England shall be the proper law of this Contract and in particular (but not so as to derogate from the generality of the foregoing) the provisions of the Arbitration Acts 1950 (notwithstanding anything in S.34 thereof) to 1979 shall apply to any arbitration under this Contract wherever the same, or any part of it, shall be conducted.

41.8 If before making his final award the Arbitrator dies or otherwise ceases to act as the Arbitrator, the Employer and the Contractor shall forthwith appoint a further Arbitrator, or, upon failure so to appoint within 14 days of any such death or cessation, then either the Employer or the Contractor may request the person named in the Appendix to appoint such further Arbitrator. Provided that no such further Arbitrator shall be entitled to disregard any direction of the previous Arbitrator or to vary or revise any award of the previous Arbitrator except to the extent that the previous Arbitrator had power so to do under the JCT Arbitration Rules and/or with the agreement of the parties and/or by the operation of law.

A footnote attached to clause 41.7 suggests that where the parties do not wish the proper law of the contract to be the law of England appropriate amendments to the clause should be made. Where the works are situated in Scotland the forms issued by the Scottish Building Contract Committee should be used. The provisions of the Arbitration Acts 1950, 1975 and 1979 do not extend to Scotland and Northern Ireland.

It is apparent from what has been said above that the arbitrator, whether appointed by the courts or as the result of conditions included in a specific contract, has considerable freedom to conduct proceedings.

Joint arbitration Construction projects frequently involve a considerable number of contractors, sub-contractors and suppliers and a dispute may raise issues which are substantially the same as issues raised in a related dispute. In this event the disputants may agree to refer the related disputes to the same arbitrator.

JCT 80 clause 41.2.1 makes provision for such an arrangement. It reads:

41.2 .1 Provided that if the dispute or difference to be referred to arbitration under this Contract raises issues which are substantially the same as or connected with issues raised in a related dispute between:
the Employer and Nominated Sub-Contractor under Agreement NSC/2 or NSC/2a as applicable, or
the Contractor and any Nominated Sub-Contractor under Sub-contract NSC/4 or NSC/4a as applicable, or
the Contractor and/or the Employer and any Nominated Supplier whose contract of sale with the Contractor provides for the matters referred to in clause 36.4.8.2,
and if the related dispute has already been referred for determination to an Arbitrator, the Employer and the Contractor hereby agree that the dispute or difference under this Contract shall be referred to the Arbitrator appointed to determine the related dispute;
that the JCT Arbitration Rules applicable to the related dispute shall apply to the dispute under this Contract;
that such Arbitrator shall have power to make such directions and all necessary awards in the same way as if the procedure of the High Court as to joining one or more defendants or joining co-defendants or third parties was available to the parties and to him; and
that the agreement and consent referred to in clause 41.6 on appeals or

applications to the High Court on any question of law shall apply to any question of law arising out of the awards of such Arbitrator in respect of all related disputes referred to him or arising in the course of the reference of all the related disputes referred to him;

41.2 .2 save that the Employer or the Contractor may require the dispute or difference under this Contract to be referred to a different Arbitrator (to be appointed under this Contract) if either of them reasonably considers that the Arbitrator appointed to determine the related dispute is not appropriately qualified to determine the dispute or difference under this Contract.

41.2 .3 Clauses 41.2.1 to 41.2.2 shall apply unless in the Appendix the words 'clauses 41.2.1 and 41.2.2 apply' have been deleted.

NSC/2, NSC/4 and DOM sub-contract conditions contain clauses which complement JCT 80 clause 41.2.1.

Clause 41.2.1 requires the disputes or differences to be 'substantially the same or connected with issues raised in a related dispute' if they are to be referred to the same arbitrator. There must be agreement between the disputants to the *joining* of the parties.

It is apparent that both main and sub-contract conditions provide for joint arbitration but JCT 80, clause 41.2.2 allows either employer or contractor to require the dispute to be referred to a different arbitrator if either of them reasonably considers that the arbitrator appointed to determine the related dispute is not appropriately qualified to determine the dispute under the contract.

There is provision in the appendix to JCT 80 for deletion of clauses 41.2.1 and 41.2.2 if the parties to the contract so wish. In this event joint arbitration would not be a condition of contract.

Advantages of joint arbitration are time-saving and consistency of award.

The powers of the arbitrator are contained in sections 1 and 12 of the 1950 Act. Section 1 states that:

'the authority of an arbitrator or umpire appointed by or by virtue of an arbitration agreement shall, unless a contrary intention is expressed in the agreement, be irrevocable except by leave of the High Court or a judge thereof'.

Bringing the dispute to court Having, above, discussed the statutory and contractual framework within which the arbitral process is conducted, the reference to arbitration and the powers of the arbitrator it is apposite to consider the practical aspects of bringing a case to a hearing.

The procedures proposed in the JCT Arbitration Rules are adopted in the following study.

Legal representation As a 'difference' (of opinion) develops into a dispute the disputants frequently find a need for legal advice on the strength of their case and for someone familiar with the procedures involved in steps 1 and 2.

Major firms often have in-house lawyers but small firms frequently find that the solicitor their firm 'grew up with' is inexperienced in dispute proceedings.

In such circumstances a disputant should consider employing one of the law firms which specialise in disputes. The 'general practice' solicitor will frequently advise his client to consult a specialist. Alternatively, the professional institutions and trade confederations should be able to advise on the names of law firms with expertise in construction disputes.

The demarcation line between solicitor and barrister is not as clear cut as in the past but in general terms solicitors initiate and carry out procedures and barristers give opinions on the strength (or weakness!) of a case.

Some specialist law firms comprise both solicitors and barristers.

Where, below, reference is made to action by one or other of the parties it should be appreciated that the action might, in practice, be taken by a solicitor on behalf of the parties.

One of the advantages claimed for arbitration is that it provides a means of settling disputes of a technical nature in an informal manner. This advantage tends to disappear if the parties engage lawyers to present their case. Furthermore, their fees add considerably to the cost of the proceedings. There has been a significant increase in legal representation in the arbitral process in recent years.

When giving his initial directions the arbitrator will usually order that if either party intends to be represented by counsel due notice shall be given to the other. This is to avoid one party being at a disadvantage by arriving at the hearing prepared to present his own case and finding himself subjected to examination by hostile counsel whilst lacking the skill to examine the other party as searchingly as he may himself have been examined.

Few arbitrators would dispute that the employment of counsel tends to add to the length of the hearing but many would say that experienced counsel make the conduct of the hearing easier for all concerned.

Service of documents Rule 3.1 seeks to identify the addressee which will be used for purposes of serving statements, documents and notices. Each party is required to notify the other party and the arbitrator of the address which is to be used.

The methods of service to be used are:

- by actual delivery to the other party, or
- by first class post, or
- where a Fax number has previously been given to the sending party, by Fax (facsimile transmission) to that number.

Where service is by Fax, for record purposes the statement, document or notice served by Fax must forthwith be sent by first class post or actually delivered (rule 3.2).

Rules 3.3 to 3.5 are concerned with proof of service and the contents of statements of case or of counterclaim.

They read:

3.3 Subject to proof to the contrary service shall be deemed to have been effected for the purpose of these Rules upon actual delivery or two days, excluding Saturdays and Sundays, after the date of posting or upon the facsimile transmission having been effected.

3.4 Any statement referred to in these rules shall:

- be in writing
- set out the factual and legal basis relied upon, and
- be served upon the other party and a copy sent to the Arbitrator.

3.5 Without prejudice to any award in respect of general damages any statement of case or of counterclaim shall so far as practicable specify the remedy which the party seeks and where a monetary sum is being sought the amount in respect of each and every head of claim.

Preliminary meeting Rule 4 of the rules requires the arbitrator to hold the preliminary meeting not later than 21 days from the Notification Date with the parties 'at such place and on such day and at such time' as he directs (rule 4.1).

The arbitrator and the parties may, however, agree that no preliminary meeting be held in which event, not later than 21 days from the Notification Date, the parties must decide which of the three procedures contained in rules 5, 6 and 7 will apply to the conduct of the arbitration (rule 4.2).

With or without a preliminary meeting, if rule 4 is followed, the procedure for the conduct of the arbitration should be agreed within 21 days of the Notification Date.

Arbitrators who are concerned with helping the parties to resolve their dispute rather than with prolonging the process suggest that the preliminary meeting provides an opportunity to bring the parties together informally and, hopefully, reach agreement. Resolution at this stage is by no means exceptional and with this in mind some arbitrators prefer to hold the preliminary meeting without legal representatives present.

Failing resolution, the preliminary meeting enables the arbitrator to meet the parties together, for the first time, to hear their statements and give *directions*. Indeed, the preliminary meeting is sometimes known as a *meeting for directions*.

Directions The arbitrator will normally give an *order for directions* at the preliminary meeting in which he orders the procedures, the arrangements and the timetable to be followed. Standard agenda should be used to ensure that all aspects are dealt with.

The arbitrator must take care to meet or communicate with both parties equally. Correspondence with one party must be copied to the other so that there may be no question of partiality. The arbitrator should not meet one party without the other being present.

Typically, the arbitrators' directions will be contained in an *order* for which the format will be:

IN THE MATTER OF THE ARBITRATION ACTS 1950 TO 1979

and IN THE MATTER OF AN ARBITRATION

Between

(name) – Claimant

(name) – Respondent

(name) – Arbitrator

ORDER

Upon representations made on behalf of the Parties at a further Preliminary Meeting held at (address) on 15 April, it is hereby **ORDERED** that the Parties shall comply timeously with my Directions numbered 1 to 19 inclusive annexed hereto.

Liberty to apply

Costs to be costs in the Reference

This Order is made at: (address)
16 April

(name)
Arbitrator

To Solicitors for the Claimant

(name)
(address)

To Solicitors for the Respondent

(name)
(address)

Directions for a traditional arbitration procedure include a programme for delivery of pleadings, the exchange of documents, provisions for experts' statements, witnesses' statements and the hearing. The following directions would be appropriate for a medium-sized case:

16 April

ORDER NO.	MATTERS ORDERED
1	There shall be pleadings in this Reference.
2	The Claimant shall deliver to the Respondent the Points of Claim on or before 13 May.
3	The Respondent shall deliver to the Claimant the Points of Defence and Counterclaim (if any) within ten weeks after receipt of the Points of Claim and in any case not later than 22 July.
4	The Claimant shall deliver to the Respondent a Reply to the Points of Defence and a Defence to the Counterclaim (if any) within four weeks after receipt of the Points of Defence and Counterclaim (if any) and in any case not later than 19 August.
5	Any request for further and/or Better Particulars of any pleading shall be delivered not later than 14 days after receipt of the relevant document, and a Reply shall be provided within a further 14 days.
6	Pleadings shall close not later than 19 August.
7	The Parties shall exchange lists of documents relating to the matters in dispute in this Reference, that have been or are in their power or possession. Lists shall be exchanged not later than close of pleadings. Discovery by inspection shall be granted within 6 weeks of close of pleadings and shall be completed not later than 30 September.
8	The Claimant shall prepare a draft bundle of documents and provide a copy thereof to the Respondent on or before 7 October. The Respondent shall satisfy the Claimant on or before 14 October, of any additional documents required to be included in the bundle. The Claimant shall complete the indexed and paginated bundle not later than 21 October.
9	Experts shall be limited to one on each side. Experts' reports shall be exchanged, for which purpose they shall be delivered to me in duplicate on or before 28 October.
10	Experts shall meet within 2 weeks after exchange of their reports in order to agree figures as figures so far as may be possible, to agree and identify all drawings, sketches, photographs and the like, and to seek to limit and identify any areas of disagreement between them.
11	Experts shall each be at liberty to prepare a commentary upon the other's report, which commentary (if any) shall be exchanged, in the manner described above, on 28 October.
12	Witnesses as to fact shall prepare sworn statements of their evidence, which statements shall be delivered by each Party upon the other of them on or before 28 October.
13	There shall be an oral Hearing in the Reference, at which evidence shall be taken under oath or upon affirmation. The sworn statements of Witnesses as to fact shall be taken as examination in chief: Experts' reports and written commentaries shall be taken as examination in chief.
14	The Hearing shall be held at (address) on 16 November commencing at 9.30 am with 17 November also reserved. The Claimant shall make reservations of a suitable room for the Hearing and of a small retiring room for my use.
15	Both Parties have notified me of their intention to be represented by Counsel.
16	The Parties have not entered into an exclusion agreement.
17	The Respondent has applied for my further or Final Award to be given in writing and with reasons.

PROCEDURES LEADING TO HEARING – IN MORE DETAIL

16 April *continued*

ORDER NO.	MATTERS ORDERED
18	I have advised the Parties of my preference for any argument or submission upon points of law to be reduced to writing for my examination prior to the Hearing.
19	The Parties shall communicate with me only in writing. All such communications shall bear on their face a confirmation that a copy has been sent contemporaneously to the other Party.

Orders for directions may state that the parties have 'liberty to apply'. The term means that the parties may apply to the arbitrator for an extension of time, for permission to amend pleadings or other action that may be taken during the interlocutory stages.

The Scott Schedule The Scott Schedule is a useful device to summarise a number of points of claim and counterclaim for the benefit of the parties and the arbitrator. There is no standard format but the arrangement shown in figure 4.4 enables the points to be tabled and costed.

It is said that disputants have occasionally, when their respective differences have been displayed in a Scott Schedule, realised that the financial implications were not as great as they had previously believed and reached an amicable settlement without recourse to further negotiations.

The Scott Schedule makes it possible to separate issues into categories. Technical and contractual issues might, for example, be identified and dealt with on different days thus reducing the amount of costly and frustrating waiting about for witnesses at a hearing.

The work involved in preparing the Scott Schedule makes the claimant and respondent familiar with the issues and puts them in perspective. It provides another point in the arbitral process at which amicable resolution of the dispute might be reached.

There is, then, reason to suggest that a Scott Schedule is prepared at an early

Item	Claim	Claimant's Case Remedy	Amount	Respondent's Case Defence	Amount	Notes	Arbitrator's Summary Award Clmt	Award Dfdt
1	Dfdt delayed completion by 6 weeks	Payment of damages at £2,200/week	£13,200	Lack of information prevented progression and led to wiring faults	–			
2	Liquidated damages arising from delay	Payment of liquidated damages	£4,600	See 1 above	–			
3	Dfdt damaged wall panels repairing wiring	Payment of cost of new panels	£2,700	Claimant has not proved damage was entirely by dfdt's man and see 1, above. Accepts partial responsibility say	£1,350			

4.4 The Scott Schedule of items in dispute

stage in the negotiations. Without submission to such a discipline it is possible for disputants to become so preoccupied with the details of the points of claim and counterclaim that the financial implications are overlooked.

Some arbitrators order, at the preliminary meeting, that the parties prepare a Scott Schedule.

The Scott Schedule in figure 4.4 assumes a dispute between a contractor and sub-contractor. The contractor is the claimant.

Pleadings 'Pleadings' is the collective noun for the points of claim, defence, counterclaim, etc. A programme for the preparation and presentation of the pleadings for a medium sized case is contained in the section concerned with directions, above.

Points of claim are, typically, delivered in response to a direction from the arbitrator. They include short sentences identifying the claimant and respondent and their interests, eg 'the Claimant is a company specialising in electrical installation and the Respondent is a building construction company'.

A paragraph identifies the nature of the agreement and relevant contract documents followed by another which refers to the clause(s) in the conditions of contract under which the dispute has arisen and under which the claim is made. The obligations of the parties under the contract are identified and the relevant failure(s) stated.

The *Particulars* (of claim) are, typically, in a claim for loss and expense, a list of headings with amounts against them which are totaled. The headings may be those used by the claimant in a detailed claim. See section 2.3.

It will be stated that the claimant claims the total and interest thereon.

The notice is served by the solicitor, assuming the parties have opted for legal representation.

Further and better particulars of the points of claim may be requested by the respondent. The request refers to specific items in the points of claim on which further and better particulars are sought, for example, details of items of work alleged to have been completed after the completion date, the precise effect alleged to flow as a result of the matters complained of, etc. Particulars will be sought of precisely how the sum is calculated and made up and to which items of complaint the amounts relate.

If the claimant is applying for reimbursement of loss and expense the request probably seeks the details contained in the original details of claim.

The claimant should find no difficulty replying to the request for further and better particulars if his claim has been properly documented, quantified and costed.

Further and better particulars may, therefore, be a lengthy document.

Counterclaim, defence, etc The format for a counterclaim, defence and any other pleadings is similar to those described above.

Examination of property by parties Either party may examine the property which is the subject of dispute. The arbitrator may be asked to order a party who is reluctant to allow the other party access to property to make an order for disclosure and/or inspection.

The term 'property' includes everything from a single component to a whole building complex.

Inspection by Arbitrator Arbitration rule 8 sets out the arbitrator's position regarding inspection.

8.1   The Arbitrator may inspect any relevant work, goods or materials whether on the site or elsewhere. Such inspection should not be treated as a hearing of the dispute.

8.2   Where under rule 8.1 the Arbitrator has decided that he will inspect:

where rule 5 applies, as soon as the parties have served all their written statements or the last of the times for such service allowed by these Rules has expired the Arbitrator shall fix a date not more than 10 days in advance for his inspection and shall inform the parties of the date and time selected; where rule 6 or rule 7 applies, the Arbitrator shall fix a date for his inspection and shall inform the parties of the date and time selected.

8.3 .1 The Arbitrator may require the Claimant or the Respondent, or a person appointed on behalf of either of them, to attend the inspection solely for the purpose of identifying relevant work, goods or materials.

.2 No other person may attend the Arbitrator's inspection unless the Arbitrator shall otherwise direct.

Rule 8.3 illustrates the way in which the arbitrator must act 'evenhandedly' with the parties. Arbitrators frequently ask both parties to be in attendance when making an inspection.

Discovery After the pleadings (the claim, counterclaim, further and better particulars, etc) have been agreed the parties prepare a list of all the documents on which they intend to rely.

These documents are scheduled and must be made available for all concerned to inspect. The documents may range from agreements, conditions of contract, specifications, relevant drawings, etc, and may include tenders from suppliers, sub-contractors, price analyses and similar items if the dispute is concerned with matters of cost.

This procedure is known as *discovery*. Should either party have reason to believe that the other has not disclosed all relevant documents he may serve a notice requiring an affidavit verifying the list or seeking a statement regarding the whereabouts of a particular document which is known to have existed but which has not been disclosed. Such a situation may arise if a disclosed letter refers to another which has not been made available.

If the parties fail to reach agreement regarding discovery between themselves, either may apply to the arbitrator to order the other to disclose the document in question. If the order fails to disclose the document there may be grounds for an application to the High Court for an *order* for discovery.

Discovery frequently takes considerable time and the cost of copying documents for even a small case may run into thousands of pounds. Lawyers tend to play safe and require that all documents in any way connected with the dispute are discovered. There are reports that lists of documents for discovery for a major dispute make reference to all the documents on a particular shelf or even all the documents held in a particular room!

A disadvantage of the JCT Rules is that each party presents all the documents on which he relies with his statement of case and there is no real provision for general discovery, only specific discovery of particular documents. The rules allow a party who wishes to disguise the fact that he has documents harmful to his case to do so legally and properly.

Privileged documents Correspondence and communication between clients and their legal advisers and documents prepared with a view to litigation and/or arbitration are generally *privileged* and need not be disclosed. Offers to settle, where these have been marked 'without prejudice', are also privileged.

Other than the above examples, claims for privilege are unlikely to be upheld.

In *Mitchell Construction Kinnear Moodie Group* v *East Anglia Regional Hospital Board* (1971) the contractor was ordered to disclose files concerned with his employees although the contractor agreed that the files were confidential. It was held that the sole issue was of relevance. Further reference to privilege is discussed under the next heading.

The section, below, concerned with 'offers to settle' provides a further example of privilege in connection with Without Prejudice documents written in the course of negotiations of a settlement.

Agreement of figures and facts When a dispute is concerned with loss and expense, and/or delay, in order to reduce the differences between the figures of the respective parties before putting them to the arbitrator, it is usual for the representative of the parties to meet and explore common ground. It is unlikely that they will reach complete agreement but if they are able, at least, to reach agreement within a range of, say £3x and £4x time and expense of the hearing may be reduced. Such meetings have on occasions led to settlement being reached without the need for reference to a hearing.

The 'representatives' referred to above may be witnesses as to fact or experts. They would more usually be independent experts but they might also be persons in the employment of the parties.

Where the dispute is concerned with loss and expense consultant quantity surveyors are frequently engaged by each of the parties as experts. Arbitrator's order No. 10, above, provides an example of such an order. An agreement reached by the experts should be presented as a written report, or privilege is attached to their discussions and it cannot be used as evidence (*Carnell Computer Technology Ltd v Unipart Group Ltd* (1988)).

Evidence The evidence involved in construction disputes may be considered under the headings:

– documentary
– oral, and
– real evidence.

It may be further categorised as evidence of fact or opinion.

Documentary evidence comprises correspondence, tenders, forms of contract, specifications, drawings, photographs, diaries, programme charts, etc, etc. The majority of evidence used in construction dispute resolution is documentary.

Oral evidence is given during the hearing under oath by affirmation. Reference is made in section 4.11 to the arbitrator's powers to administer oaths or take affirmations.

A witness who knowingly makes a false statement under oath may be punished by fine and/or imprisonment.

A reluctant witness may be compelled to give evidence at a hearing by means of a writ of *subpoena ad testificandum*.

Where evidence is required from a person who is unable to attend the court it may be given in the form of an affidavit. Such sworn statements provide sound evidence but as the person is not available for cross-examination such statements do not carry as much weight as oral evidence given in court.

Real evidence is a material object which may be brought into court or examined in situ. A broken flue-pipe which caused a fire, for example, might be examined in either location.

Evidence of fact may be given by any reliable person. A child's evidence has been accepted in court. No experience or professional qualifications are necessary if the evidence is, for example, concerned with damage seen to be caused by a machine which ran into the side of a building or a ceiling height which does not comply with a dimensioned drawing.

Evidence as to opinion is the province of the expert witness who should be suitably qualified or experienced.

The expert witness is subject to the provisions of the Civil Evidence Act 1972. His role is to assist the arbitrator to come to a correct decision on a particular point. This evidence should be concerned with truth rather than with presentation of the case of the party who called him. When, as is not unusual, one hears experts called by the parties whose evidence is diametrically opposed it is difficult to believe that both are concerned with assisting the arbitrator rather than with advancing the cause of the party who called them.

Concern of the court in this respect was expressed by Lord Wilberforce in *Whitehouse v Jordan* (1981).

'It is necessary that expert evidence presented to the court should be and should be seen to be the independent product of the expert, uninfluenced as to form and content by the exigencies of litigation. To the extent that it is not, the evidence is likely to be not only incorrect but self-defeating.'

Arbitrators and judges have been known to comment that expert witnesses who appear frequently in their courts espousing different causes on different occasions tend to lose credibility in their eyes.

Hearsay evidence is a statement alleged to have been made by another person. Such evidence is, in general, not admissible unless the other person is the opposing party (or a person representing him) or unless the statement was made in the presence of the opposing party.

Wherever possible the person making the statement should be called to give evidence himself. Hearsay evidence usually carries less weight than evidence given in court even though the Civil Evidence Act 1968 has relaxed and extended some rules regarding hearsay evidence.

Burden of proof of evidence rests with the party making the assertion. When proof has been provided the burden moves to the other party to disprove the assertion or to show that it is not material (relevant).

Preponderance of probability It is often difficult to provide absolute proof and in the absence of absolute proof the arbitrator must decide on the *preponderance of probability*. In effect, he weighs the evidence and finds in favour of the party whose evidence is weightier.

The admissibility of evidence rests, largely, with the arbitrator. Where a party does not have legal representation the arbitrator may draw the attention of the party to defective evidence being presented by the other and invite objection. The arbitrator should not refuse to hear admissible evidence but he may encourage a party to brevity who is, in his view, labouring a point on the grounds that he has heard enough on a particular point.

Offers to settle References are made elsewhere in this section to the majority of disputes being settled before they reach a hearing. One reason for settlement is that the cost of an arbitration hearing may be high and there is always the risk that the case will be lost. Even if it succeeds the winner is unlikely to recover all his costs.

Crowter[4] gives an example of an arbitration where the claim is for £50,000. The respondent offers £20,000 to settle which is rejected. The arbitration proceeds to an award, where the arbitrator only awards the claimant £10,000. Clearly, the claimant should have accepted the offer; continuing the arbitration was a waste of time and money. Therefore, despite the fact that the claimant 'won' in that he was awarded £10,000 the respondent will be entitled to his costs from the time when the claimant should have accepted the offer of £20,000.

It follows that offers to settle are tactically important.

Four types of offer may be identified:

– open offers
– without prejudice offers
– sealed offers, and
– Calderbank offer, or offer 'without prejudice save as to costs'.

Open offers Any party is entitled to make an open offer, ie one that may freely be brought to the attention of the arbitrator by either party. Such an offer should always be made in writing. The great disadvantage with the open offer is that the arbitrator is likely to be embarrassed by seeing it. It is very difficult for an arbitrator not to be influenced when he knows how much one party is prepared to offer to settle. For this reason open offers are unusual and disliked by arbitrators and the parties' legal advisers.

Without prejudice offers Offers to settle or compromise a dispute, either marked 'without prejudice' or even when not marked, clearly made in the process of attempting to compromise a dispute, will for public policy reasons, not be admissible in evidence in arbitration proceedings. The arbitrator will be totally ignorant of such an offer which cannot be brought to his attention at any stage of the arbitral process. In other words, the arbitrator's award of costs will not be influenced in any way by the offer, even it is was for a sum in excess of the award.

A most interesting case concerning the continued privilege attaching to Without Prejudice documents written in the course of negotiations of a settlement, even in other connected litigation, was decided by the House of Lords in November 1988. The case was *Rush and Tompkins v GLC* and others Times Law Report, 8 November 1988, 43 BLR 1.

It was held that the 'Without Prejudice' rule governed the admissibility of evidence and was founded on the public policy of encouraging litigants to settle their differences rather than litigate them to the finish. The rule applied to exclude all negotiations genuinely aimed at settlement, whether oral or in writing, from being given in evidence.

The application of the rule did not depend on the use of the words 'Without Prejudice', and if it was clear from the surrounding circumstances that parties were seeking to compromise an action, evidence of the contents of their negotiations would not, as a general rule, be admissible at the trial and could not be used to establish an admission.

It was also held that under the 'Without Prejudice' rule, admissions made in negotiations genuinely aimed at settlement were inadmissible in evidence in any subsequent litigation connected with the same subject matter.

Furthermore, admissions made to reach a settlement with a different party within the same litigation will also be inadmissible. The same rule was to be applied to protect such settlement negotiations from production on discovery to other parties in the same litigation.

Accordingly where a plaintiff brought an action against two defendants, 'Without Prejudice' correspondence, which had passed between the plaintiff and one of the defendants and had resulted in a settlement of their claims, was protected from disclosure to the other defendant against whom the action continued. No doubt the same principles apply in arbitration.

Sealed offers The sealed offer was for many years the arbitral equivalent to a payment into court in litigation. The principle in litigation is that where the defendant to a money claim makes a payment into court which equals or exceeds the amount ultimately awarded to the plaintiff, the defendant is entitled to be awarded his costs incurred after the plaintiff received notice of the payment-in. The payment-in is completely hidden from the judge until after he has made his judgement on all matters with the exception of costs, at which point his attention is drawn to it. Court procedure lends itself to the payment-in system as judgements are invariably given orally even if they are reserved and an opportunity therefore arises for counsel to address the Judge on the question of costs after the judgement on liability and quantum has been handed down.

It is rare for such opportunities to arise in arbitration. Therefore, the procedure of sealed offers came into being. A written offer is made by one of the parties on a 'Without Prejudice' basis, which is subsequently rejected. The party making the offer then informs the other party that a copy of the offer and usually the letter of rejection will be placed in a sealed envelope and handed to the arbitrator at the end of the hearing with a request that the does not open it until he has made his award on all matters except costs. Thus in theory, the arbitrator completes his substantive award in ignorance of the amount of the offer to settle (although he knows there is such an offer in existence), and then and only then opens the envelope before he awards costs and their apportionment.

The sealed offer has three main disadvantages:

– the arbitrator knows that an offer has been made to settle and probably, by

inference, who has made the offer. It is not impossible that this could influence him in coming to his award
- a sealed offer, unlike a payment into court, is not backed by cash; but this applies to all offers to settle in arbitration
- there is normally no opportunity for the arbitrator to be addressed either orally or in writing on the effect that the sealed offer should have on his award for costs, or on the nature of the sealed offer itself. For example, one party could say that he had insufficient time to consider the sealed offer and it was therefore invalid; or alternatively a party could say that the offer included costs and he had no way of knowing what was the capital sum offered excluding costs and therefore he was entitled to reject it in any event.

The first disadvantage is sometimes overcome by the respondent presenting the arbitrator with an envelope at the end of the hearing in any event; some arbitrators even insist on this procedure. The envelope might contain a sealed offer or it might contain a statement that no sealed offer had been made. Providing the arbitrator knows that either of these alternatives might apply, he will not be tempted to be influenced either way.

Calderbank offer 'without prejudice save as to costs' This is essentially an improvement and sophistication of the sealed offer procedure. It is named after the case *Calderbank v Calderbank* (1976) Fam 93 where the procedure was first approved by the courts.

An offer to settle is made marked 'without prejudice save as to costs'. This means that such an offer remains privileged, ie cannot be disclosed to the arbitrator, until the arbitrator comes to consider costs when the document containing the offer is put before the arbitrator as additional evidence upon which both parties will have a right to further address him.

The procedure means that the arbitrator will not make a final award immediately, but will publish an interim award dealing with liability and quantum; reserving his award on costs to a final award. Normally the parties will be invited to try to agree an order for costs based on the interim award, but if no agreement is forthcoming the hearing will reconvene for the taking of further evidence on costs, after which the arbitrator will make his final award.

It is the essence of a Calderbank offer that it is not put in evidence until after the substantive award has been made. This is an exception to the general rule that once an award has been published no other evidence can be admitted.

The Calderbank procedure completely overcomes the problem of the arbitrator being influenced by the known existence of an offer. Further it gives the parties the right to argue their case on costs before the arbitrator who is then in full possession of the history of attempts to settle. Arbitrators may be well advised to bring this procedure to the notice of parties at preliminary meetings.

The Calderbank offer was first approved in a court action, but first the courts and then an amendment to the Rules of the Supreme Court re-affirmed the pre-eminence of the payment-in system with the Calderbank procedure only being permissible where payment-in is not appropriate. No such restrictions apply to arbitration.

Written offers 'Without prejudice save as to costs' have been specifically approved by inclusion in the Rules of the Supreme Court (Amendment) 1986 – Order 22 rule 14.

The importance essentials of the Calderbank offer are that it is in writing and that the words 'Without prejudice save as to costs' are prominently displayed on it.

Important aspects of the offer to settle:

- Interest An offer should include interest up to the date of the offer.
- Costs   Offers should always be for £X plus costs. An offer which includes costs as part of the sum offered should not be considered in the context of an order for costs. The arbitrator has no knowledge of the amount of costs at the date of the offer. He is therefore unable to make the

vital comparison between the award and the offer exclusive of costs. Offers including costs are very common, but any party wishing to protect its interests in costs should avoid the temptation to make this type of offer.

How long should the offer remain open? Any offer to settle should be kept open long enough for the party receiving it to be able to consider carefully whether it should be accepted. It is suggested that the offer should be held open for 21 days providing that period can elapse before the commencement of the hearing. Within 21 days of the hearing the offer should be held open until the first day of the hearing. It is also not inconceivable that an offer will be made during the course of the hearing. Because the hearing involves the most intensive expenditure of costs and both parties' minds are concentrated totally on the matters in dispute, such an offer will normally be held open for no longer than 24 hours.

Tactically an offer should always have an expiry date. There was a school of thought that offers should remain open until the end of the arbitration proceedings, but this lacks the tactical advantage of pressurising the claimant into an early decision, possibly before all the documents have been discovered and possibly before expert reports have been exchanged.

A claimant should think very carefully before rejecting an offer, as he is under a greatly increased risk of costs from that moment on.

Timing of the offer In large arbitrations with counsel, particularly leading counsel, a point of time arrives when a substantial brief fee becomes payable. It is particularly important to try to make the offer before that fee is paid. Some respondents will, in their offer, undertake to pay the claimant's costs up to the receipt of the offer plus only the further costs of considering the offer. In other words further preparation costs between the date of the offer and the date it lapses are excluded. It is not certain whether such a tactic would be successful if contested in court.

The effect of the offer on costs The claimant will be entitled in any event to his costs up to the time the offer was received. He will further be entitled to at least the additional costs of considering the offer. If the offer is rejected or lapses and the matter proceeds to an arbitration hearing, the arbitrator will have been asked to give an interim award dealing with liability and quantum only. In that award, if the claimant has failed to beat the offer he will be responsible for the respondent's costs from the time the offer was rejected or lapsed. If the claimant beats the offer, he is vindicated in his decision to reject the offer and to proceed, and is therefore entitled to recover his costs in total (in principle at least).

This simple view is complicated by two main difficulties, interest and counterclaims.

Interest When an arbitrator is considering his award on costs, he may have a Calderbank or other offer before him that was made many months or even years before. It is possible that although his award is more than the offer made, it might not be if nominal interest was added to the offer to bring it up to date. For example, an offer of £50,000 was made and rejected in 1987; the arbitrator awarded £55,000 in 1989; so effectively the claimant beat the offer and, following the general rule, would be entitled to his costs. However, if nominal interest was added to the 1987 offer, the 1989 equivalent would be, say, £62,000. In theory the 1989 award could also be reduced to 1987 levels, but most arbitrators will prefer to add nominal interest to the offer. The important factor is to compare like with like. Thus in the example, although at first sight the claimant has beaten the offer, in reality he has not and should have settled in 1987 for £50,000. The arbitrator will therefore award costs to the respondent from 1987 onwards.

Counterclaims Counterclaims present another problem. It must first be decided whether the counterclaim is an independent action that would stand alone if the claim did not exist, or whether it is in reality a defence to the claim and could have no independent life of its own. In the words of the old adage, is it a sword or a shield?

The general practice of arbitrators is that if the counterclaim would in itself give rise to an independent action, the costs of the claim and counterclaim will be awarded separately. For example, an arbitrator might say: 'The respondent shall pay the cost of the claim and the claimant shall pay the costs of the counterclaim'. This is known as 'a cross order for costs'. It can of course lead to problems for the parties in determining which costs apply to the claim and which to the counterclaim. If the counterclaim is in reality a defence or a shield, then an arbitrator will normally make only one order for costs.

This procedure raises many fascinating tactical possibilities with offers to settle. For example, there is nothing to stop a claimant making an offer to settle the counterclaim without mentioning the claim. Such an offer is normally of great tactical advantage to the claimant. Likewise, if the counterclaim is substantial and meritorious, the respondent may offer to settle the claim without offering to compromise the counterclaim. It is easy to imagine further possibilities. For example, in many building cases, there are a whole series of independent claims and counterclaims. It is open to the parties to make offers to settle individual claims and counterclaims either individually or collectively. The tactical advantages can be enormous.

4.7 The hearing

The *hearing* is the stage in the arbitral process which is frequently regarded as the culmination. There are no precise records of the number of cases which are referred to settlement by arbitration but which are not heard but a popular statistic is that at least 90% of the referred cases are settled out of court. But even if hearings are the exception the compilation of evidence leading yo the hearing may have been essential to settlement of the dispute.

For many years the mock arbitration hearing has provided a case for study of the application of procedures and rules discussed in section 4.6.

The case study in section 4.11 includes a synopsis, script and award used for a mock arbitration hearing. They are largely self-explanatory but some comments, of the sort which might be given by the arbitrator during the event, have been introduced where appropriate.

Frivolous and/or mythological names for the 'dramatis personae' are part of the mock hearing tradition. The substance of the points of claim has been simplified in order to concentrate on the procedures.[5]

4.8 The award

Sections 13, 14, 15, 16, 17 and 20 of the 1950 Act are concerned directly with the arbitrator's powers to make an award. Other sections are concerned with certain foreign awards.

The above sections state that:

- he has the power to make an award at any time but that the High Court may, on application from any of the parties remove him if he fails to 'use all reasonable dispatch' (section 13)
- he may make an interim award (section 14)
- he may 'order specific performance' (section 15)
- his award is to be final unless there is a contrary intention expressed in the arbitration agreement (section 16)
- he has the power to correct in an award any clerical mistake or error arising from any accidental slip or omission (section 17)
- a sum directed to be paid by an award shall, unless the award otherwise directs, carry interest as from the date of the award at the same rate as a judgment debt (section 20)

The principal amendment in the 1979 Act which is concerned with awards relates to the issue of a reasoned award (section 1).

The Acts provide a framework for drafting arbitration awards and JCT Arbitration rule 11 provides:

11.1 The Arbitrator shall only give reasons for his award where and to the extent required by either party by notice in writing to the Arbitrator with a copy to the other party.

11.2 .1 The Arbitrator may from time to time publish an interim award.
.2 If in any interim award the parties are directed to seek agreement on an amount or amounts due but such agreement is not reached by the parties within 28 days of the receipt of that award (or within such other lesser or greater period as the Arbitrator may direct) the Arbitrator shall, on the basis of such further appropriate evidence or submissions as he may require, publish a further award on the amount due in respect of any liability or liabilities set out in the interim award.

11.3 On publishing an award the Arbitrator shall simultaneously send to the parties by first class post, a notification that his award is published and of the amount of his fees and expenses (or any balance thereof if rule 9.2 has applied).

11.4 An Arbitrator's award can be taken up by either party on payment to the Arbitrator of his fees and expenses. The Arbitrator shall forthwith deliver the original award to the party who paid his fees and expenses and shall simultaneously send a certified copy of the award to the other party.

11.5 If, before an award is published, the parties agree on a settlement of the dispute the parties shall so notify the Arbitrator. The Arbitrator shall issue an order for the termination of the arbitration or, if requested by both parties and accepted by the Arbitrator, record the settlement in the form of a consent award. The Arbitrator's fees and expenses shall be paid upon notification that such order or consent award is ready for taking up and on payment thereof the Arbitrator shall be discharged and the reference to arbitration concluded.

A format for an award is given in the case study in section 4.11.

The term 'fit for counsel', below Award, Item (5), enables the parties to include counsels' fees and expenses in their bills of costs.

The award in the case study requires payment by one party to the other. Some disputes are concerned with the failure of one party and/or the other to do what they contracted to do. Under such circumstances the arbitrator may order the *specific performance* referred to in section 15 of the 1950 Act, above.

Enforcing the award Section 26 of the 1950 Act states that an award on an arbitration agreement may, by leave of the High Court or a judge thereof, be enforced in the same manner as a judgement or order to the same effect, and where leave is so given judgement may be entered in terms of the award.

Payment to trustee-stakeholder Rule 10 makes provision for payment to be made to a trustee-stakeholder in the event of the arbitrator making his award in favour of the claimant before he has published his award on all matters in a counterclaim by the respondent:

The Arbitrator upon application by the Claimant or the Respondent and after considering any representations by the parties may direct that the whole or a part of the amount so awarded shall be deposited by the Respondent with a deposit-taking bank to hold as a trustee-stakeholder (as described in Rule 10.3) pending a direction of the Arbitrator under Rule 10.2.1 or of the parties under Rule 10.2.2 or of the court under Rule 10.2.3.

The trustee-stakeholder shall hold any amount deposited as a result of a direction of the arbitrator under rule 10.1 in trust for the parties until such time as either:

.1 the Arbitrator shall direct the trustee-stakeholder (whether as a result of his award or as a result of an agreement between the parties reported to the Arbitrator or otherwise) to whom the amount deposited, including any interest accrued thereon, should be paid by the trustee-stakeholder; or

.2 if the Arbitrator is deceased or otherwise unable to issue any direction to the trustee-stakeholder under Rule 10.2.1 and the Arbitrator has not been replaced, the parties in a joint letter signed by or on behalf of each of them direct the trustee-stakeholder to whom the amount deposited, including any interest accrued thereon, should be paid by the trustee-stakeholder; or

.3 a court of competent jurisdiction gives directions.

Remission of and setting aside award
Section 22 of the 1950 Act states that in all cases the High Court or a judge may from time to time remit matters referred to the re-consideration of the arbitrator. In this event the arbitrator has three months after the date of the order of remission in which to make his award.

An arbitrator may be removed by the High Court if he has misconducted himself or the proceedings and an award which has been improperly procured may be set aside.

Misconduct includes making a procedural error.

An award may be set aside if in the opinion of the court the arbitrator has made an error of law.

What is 'a question of law'?

The case called *Pioneer Shipping Ltd v BTP Tioxide Ltd*, commonly known as 'The Nema' which was decided in the House of Lords provides guidelines. The case was heard by one of the most experienced and respected shipping arbitrators. His award was appealed and reversed in the High Court. The Court of Appeal decided the appeal should not have been given and that decision was upheld in the House of Lords.

The established principle is that where parties to a commercial contract submit their dispute to a commercial arbitrator, his commercial judgment on the issues should generally be allowed to stand and should not be subject to appeal even though it might involve a decision which is clearly one of law.

If the dispute arises under a contract devised expressly for the purposes of the parties, what is known as a 'one-off' contract, the arbitrator's commercial judgment as to what the parties must have intended the contract to mean should stand unless the judge to whom an application for leave to appeal is made considers that the arbitrator has made so obvious and blatant an error of law that the judge would take a great deal of convincing that an error had not been made.[6]

Reasoned award For arbitrations concerned with matters of quality or matters of fact a simple award without reasons being given is usually sufficient.

There may in other cases be advantages to the arbitrator stating his reasons when making his award.

Indeed, judges have made it known that they will expect arbitrators to state their reasons as a matter of normal practice where the complexity or weight of the issues is other than minimal. An arbitrator may and should refuse to give a reasoned award if he considers it would be inappropriate to do so.

Nevertheless, Gibson-Jarvie and Hawker suggest that 'the best precept for the arbitrator is probably, if in doubt, give reasons'.

A precis of their suggestions for a suitable general framework is:

– simple, straightforward, non-technical language should be used
– jargon of a trade or profession should be avoided
– the award should be as short and concise as possible in good, plain English in narrative form
– open with short account of how the contract was set up
– state the events leading up to the dispute
– state what action was taken to resolve the dispute before recourse to arbitration
– state how the arbitrator came to be appointed
– give a concise summary of the issues and contentions
– discuss the evidence if the facts are not agreed and the arbitrator's conclusions stated
– state the arbitrator's views on the appli-

cation of the relevant points of law leading to his decisions
- give his findings set out in formal language and his orders for damages, costs etc or other remedies
- address all matters referred to him and avoid addressing matters not referred to him
- give as appendices only items without which the award would be rendered incomplete
- give his reasons for saying that he sees no grounds for appeal, if that is his view.

Arguments in favour of the arbitrator giving a reasoned award are:
(i) that the recipients of the award will know 'where they went wrong' (or right) in the opinion of a third party, the arbitrator, and may profit from their experience; and
(ii) that the arbitrator will take greater care when making an award if he has to declare his reasons.

An argument against the reasoned award is that the existence of the reasons may provide the disputants with an opportunity to dispute them thus opening the way for delay. It is frequently suggested that the majority of disputants seek a quick, cheap and as-fair-as-possible resolution to their dispute and that anything which delays and adds to cost is to be discouraged.

As to how the reasons should be set out and what areas should be covered, Lord Diplock, in a speech made during the Third Reading of the Bill said:

'when reasons are given they can be very simply expressed in ordinary language and not in the rather technical form which case-stated cases now take'.[7]

4.9 Arbitrator's fees and expenses – costs

Arbitration rule 9 provides:

9.1 From the Notification Date the parties shall be jointly and severally liable to the Arbitrator for the payment of his fees and expenses.

9.2 In an arbitration which continues for more than 3 months after the Notification Date the Arbitrator shall be entitled to render fee notes at no less than 3-monthly intervals and the same shall be payable 14 days after delivery.

9.3 The Arbitrator shall, unless the parties inform him that they have otherwise agreed, include in his award his decision on the liability of the parties as between themselves for the payment of his fees and expenses and, subject to Rule 7.1.2, on the payment by one party of any costs of the other party.

9.4 The Claimant shall, unless the Respondent has previously done so, take up an award of the Arbitrator and pay his fees and expenses (or any balance thereof if Rule 9.2 has applied) within 10 days of the notification given by the Arbitrator to the parties of publication of the award as provided in Rule 11.3

Arbitrators generally state when accepting appointment their charges and any arrangements with regard to payment of fees and expenses.

As arbitrators' appointment diaries for hearings are frequently booked up well ahead and parties frequently settle 'out of court' shortly before the hearing, arbitrators may make cancellation charges in the event of the parties not proceeding to the hearing.

Costs, fees and interest Section 18 of the 1950 Act states that the costs of the reference and award shall be at the discretion of the arbitrator who may direct to whom and by whom and in what manner costs shall be paid and may tax or settle the amount of costs to be paid. He may award costs to be paid as between solicitor and client.

Any costs directed by an award to be paid shall, unless the award otherwise directs, be taxable in the High Court.

The parties cannot enter into an agreement which provides for them to pay their own costs of the reference or award unless such agreement is entered into after the dispute has arisen.

If the arbitrator does not make provision for costs in his awards the parties may, within 14 days of the publication of the award apply to the arbitrator for an order directing by and to whom costs shall be paid. The arbitrator must then amend his award to include the costs.

The basic rule with regard to costs is that they *follow the event* which means, in effect, that the successful party is awarded his costs. This rule works satisfactorily in most cases but section 4.6, offer to settle, discusses circumstances under which the rule may not produce the best solution.

A party who causes unnecessary costs should bear them regardless of the outcome of the arbitration.

Examples are given above in the section concerned with offers to settle, of the effect of an offer on costs and of the significance of interest on the arbitrator's award.

4.10 Advantages and disadvantages of arbitration

Earlier in this section passing reference has been made to the advantages and disadvantages of arbitration for the parties to the contract whose principal concerns are for settlement of their disputes as quickly and as economically as possible.

In section 4.2 the merits of alternative approaches for resolution of disputes are discussed. These alternatives may provide quick and economical means of settlement but they lack the legal enforceability of arbitration and litigation. We are concerned, below, with the advantages and disadvantages of arbitration when compared with litigation.

- Provided the parties agree on the appointment of the arbitrator, arbitration should provide a quicker means of resolution than litigation, which is dependent on the availability of a judge or official referee. Without goodwill, however, arbitration may be no quicker than litigation.
- Provided the parties avoid legal representation or keep it to a minimum, arbitration should be more economical than litigation.
- Arbitration involves submission of a dispute of a technical or commercial nature to a third party who has technical or commercial expertise. The parties to the dispute can choose the person they consider best qualified to hear it or they may rely on the person nominated in the conditions of contract to appoint.
- Arbitration is private. The award is made known only to the parties to the dispute.

The years preceding the publication of the JCT Arbitration Rules in 1988 saw arbitration becoming an unacceptably lengthy and costly process. Implementation of the rules should restore the advantages of the arbitral process as a quick and economical means of dispute settlement.

4.11 Case study
Arbitration, Lucifer v Barchester

Synopsis The action is based on the Standard Form of Contract, Local Authorities Edition with Quantities dated 1980 and Standard Method of Measurement, sixth edition.

The Barchester Borough County Council caused plans to be prepared by their staff architect for a secondary modern school and open air swimming pool. Owing to pressure of work, the quantity surveying department of the Council were unable to carry out the preparation of the bills of quantities.

In consequence of this, an outside firm of chartered surveyors, Messrs Angel Gabriel and Partners, was employed for the contract.

Trial holes were dug and eventually the bills of quantities were prepared and sent to the contractors invited to tender.

Within the bills of quantities certain conditions were laid down as to form of contract, reference to drawings, trial holes, site visits, and various descriptions and dimensions for excavation for the swimming bath, which have a bearing on the dispute.

There was no mention that the bills of quantities were not measured in accordance with the Standard Method of Measurement.

The lowest tender received was that of Messrs John Lucifer and Sons Ltd, and in due time a contract was signed and the appendix completed.

The architect named was Mr Hercules Jupiter, and the surveyor as Messrs Angel Gabriel and Partners.

In a matter of two or three weeks the contractor received a letter from the architect stating that the position of the swimming bath might be altered from that shown on the drawing.

Some three months later the contractor asked for definite instructions as to the position of the swimming bath as all other excavation had been completed and excavating plant was standing idle. After a period of three weeks or so, Variation Order No. 921 was issued in the following terms:

'Change the position of the swimming bath and construct all in accordance with drawing No. 5, in the position shown on drawing No. 104.'

Excavation was commenced immediately, and at a depth of 600 mm below the surface hard rock was encountered. At the next meeting of the contractor and quantity surveyor for the purpose of an interim certificate a quantity for rock and an extra over price was agreed and included in the certificate for the amount of rock excavated to date. Subsequently, a total quantity of rock was agreed and included in the interim certificate following completion of the excavation of the swimming bath.

Application for extensions of time were applied for and granted, the revised completion date being 1 October, but the actual date of completion was 1 December.

In July of that year Mr Jupiter, the borough architect, accepted an appointment in Moscow, and Mr Hector Pluto was appointed to succeed him.

The final account was produced in March, and submitted to Mr Pluto, who in April passed it to the borough treasurer, Mr Jason Midas.

Certain items of query were then passed back to Mr Pluto, the architect, who then passed on this letter to the borough treasurer, Mr Midas, who still adhered to his previous opinion.

The education committee was given a copy of the Final Account prepared by Messrs Angel Gabriel and Partners, together with the observations by the borough treasurer. The committee resolved that a further interim payment, lower than the amount shown to be due, be paid until such time as the treasurer and quantity surveyor resolved their differences.

The contractor completed the maintenance works to the satisfaction of the architect, Mr Pluto, and asked for his final payment. He was informed that there were differences of opinion between the treasurer and quantity surveyor, and that the education committee had instructed the architect not to issue the final certificate.

The contractor wrote to the borough architect saying that he could not accept the situation, and that unless the balance due was paid he would take the matter to arbitration.

A number of meetings between the borough treasurer and the quantity surveyor took place, and eventually a meeting was arranged between Mr Lucifer, the contractor, the quantity surveyor, the borough architect, the borough treasurer, and the town clerk.

It was eventually agreed between Mr Lucifer and the town clerk that the council would be recommended by the town clerk that an application be made for the appointment of an arbitrator, and consequently the President of the RIBA appointed Norman Royce Esq FRIBA, FCI Arb.

THE HEARING

DRAMATIS PERSONAE
Mr Jorrocks   QC for the Claimants
Mr Pratt   QC for the Respondents
Mr Lucifer   The Contractor
Mr Herald Angel   Quantity Surveyor
Mr Hector Pluto   Architect

(COMMENT: *The arbitrator gives instructions regarding the layout of the court and the level of formality to be adopted. Whether, for example, the parties to stand when addressing the court or to remain seated.*)

Jorrocks   May it please you Sir,
I appear for the claimants, Messrs John Lucifer and Sons Ltd and my friend Mr Pratt appears for the respondents who are the Mayor, Aldermen and Burgesses of the County Borough of Barchester.

The parties have entered into a contract dated 1st April (and you may consider that date to be an appropriate one). By the terms of that contract the claimants contracted to build a school for the respondents, and certain disputes have arisen out of that contract.

The contract documents are the form of contract published by the Royal Institute of British Architects for the use of local authorities where quantities form part of the contract, certain drawings and the bills of quantities all of which are in your possession.

Under the terms of that contract a Mr Hercules Jupiter was named as architect and the firm of Messrs Angel Gabriel and Partners were named as quantity surveyors.

During the course of the contract Mr Jupiter left the employment of the respondents and was succeeded by Mr Hector Pluto.

By clause 2.2 of the conditions of contract it is stated that the bills of quantities unless otherwise expressly stated shall be deemed to have been prepared in accordance with the principles of the Standard Method of Measurement of Building Works, a document which you, Sir, no doubt have at your finger tips. It is nowhere expressly stated within the contract that the bills of quantities have not been prepared in that manner.

During the progress of the contract the work was varied from time to time by means of written Variation Orders under clause 13 of the Conditions of Contract. One of those Variation Orders, No. 921, provided for a change in the position of a swimming bath which was included in the contract. That change involved the contractor in excavation in rock, which was not contemplated within the contract, and section 'D' Excavation and Earthwork of the Standard Method of Measurement, referred to, provides that excavation in rock shall be given separately or may be described as extra over the various classes of excavation.

Rock was not so described in the bills of quantities and it is our case that we are entitled to be paid for such excavation in rock as has occurred.

The whole of such excavation in rock arises out of Variation Order No. 921. Clause 13 of the conditions provides that the surveyor shall measure and value variations. That excavation in rock was measured and agreed with the claimants, but was subsequently rejected by the respondents. We say that they have not power to exercise such rejection.

A further dispute arose out of a delay in the completion of the works. The conditions of contract provided in clause 23 that the works should be completed by a certain date subject to extensions of time which can be given under clause 25 of those conditions. Such extensions of time were given, but unfortunately the contract was not completed within the extended completion date and the respondents claim to make reductions in the contract sum for that clause.

It is also a part of our case that by their failure to pay the sums to which the claimants are entitled the employer has interfered in such a manner as to cause loss to the claimants in respect of which we are entitled to recover damages.

I have made my opening as brief as possible and I trust that the case will develop as the witnesses are called. I propose to call two witnesses.

Mr Pratt I have one witness to call.

Jorrocks With your permission, Sir, and that of my friend I propose to do a good deal of leading in order to save as much of your valuable time as possible. Possibly if my friend feels that I am going too far in this matter he will intervene.

(COMMENT: *The extent to which the person carrying out the examination and cross-examination of a witness may 'lead' him is largely a matter for the discretion of the arbitrator who would be influenced by objections from the opposing counsel, if any. Arbitration hearings are normally more informal than those in some other courts. In the absence of legal representation the parties and the arbitrator are at liberty to question the 'other' party when he has given evidence.*

Before each witness gives evidence the arbitrator most administer the oath or require the witness to affirm. Should a witness be recalled he is reminded that he is still 'on oath'.)

Mr Pratt If you please.

Mr J Call Mr Lucifer.

(Jorrocks – Lucifer, In-Chief)

J Are you Mr John Lucifer, a director of Messrs John Lucifer and Sons Ltd, building contractors of Barchester, and have you been engaged in the building trade for 15 years?

L Yes.

J Before you prepared your tender for this contract, did you visit the site and make inspection of four trial holes?

L That I did. They were covered by long grass and when we were looking for them I slipped

J   on some wet clay and rolled into the last hole.
J   What did you say as you were rolling into this hole?
L   I said 'Soil, sand and clay'.
J   No rock?
L   Not a trace in any hole.
J   When the work on the site started, did you receive a letter?
L   We got one from the architect telling us to hold up the swimming bath.
J   When was that?
L   May.
J   Then what?
L   By August we had finished all excavation and I asked the architect about the swimming bath.
J   What happened then?
L   We got a Variation Order No. 921 telling us where we could put the swimming bath.
J   Did it say 'Alter the position of the swimming bath and construct all in accordance with contract drawing No. 5 in the position shown on drawing No. 104?'
L   Yes.
J   And you put it there?
L   We did.
J   What did you find when you started excavation?
L   Hard rock 600 mm below the surface.
J   Did you discuss it with anybody?
L   I told Mr Jupiter, the architect, that we should want an extra for it.
J   What did he say?
L   Told me to see Mr Angel.
J   Did you?
L   Yes. Mr Angel came on the site to prepare the next Interim Certificate. I told him we wanted paying extra for rock.
J   What did he say?
L   He told me that was his business, but he included a payment for rock in the next certificate.
J   Were you satisfied?
L   Oh yes. You can trust Mr Angel. He measured it up and put it in the Final Account.
J   In what amount?
L   £12,000.00

J   What happened when the contract was finished?
L   We got a Final Account from Mr Angel, and we agreed it and we sent it to the architect.
J   What happened then?
L   We got a certificate for £10,000.00
J   Was that enough?
L   No, we should have had £23,062.00
J   What did you do?
L   Went to see the new architect, Mr Pluto. He told us the Borough Treasurer had knocked £13,062.00 off the Final Account, £62 for mistakes, £12,000.00 for rock and £1,000.00 for damages.
J   What did you say?
L   I told him it wasn't good enough. He could have the £62 but we wanted the rest.
J   Did he agree to pay you the rest?
L   No. He said the Education Committee had passed a resolution saying we could have no more till the Borough Treasurer and Mr Angel had reached agreement.
J   What then?
L   I told him we could pass resolutions as well as the Education Committee and we would pass a resolution saying we should have our money.
J   Did you get your money?
L   No. We had a meeting with the whole bunch of them but I could get no satisfaction so we started this arbitration.
J   All things come to him who waits, even justice.

(Pratt – Lucifer, Cross-examination)

P   Mr Lucifer, did you read the bills of quantities before you prepared your tender?
L   Of course.
P   Did you read this – 'Include for removing any naturally occurring stone or rock that may be encountered in the course of the excavation'?
L   Yes.
P   And did you include for removing any naturally occurring stone or rock that may be encountered in the course of the excavation?
L   Yes.
P   Then why do you claim to be paid for excavation in rock?

L That doesn't mean excavation in rock. It means any odds and ends of stones you might meet when you start digging.
P That was your interpretation of the term?
L It was the common sense interpretation.
P Did you also read this – 'excavate in any material not exceeding 1.5 m deep for swimming bath and get out'?
L Yes.
P What is any material?
L Well, it's not rock.
P What is it?
L Anything but rock.
P Why not rock?
L Rock's always measured.
P And you think it should always be so measured in spite of the extracts you have just read from the bills of quantities?
L If we weren't entitled to be paid for it you can be certain that Mr Angel would have made certain it would never have gone into the Final Account.
P Do you know that there are arithmetical errors amounting to £62 in the Final Account?
L I'm not going to argue about £62. You can have that.
P I am afraid that I shall want a good deal more than £62. The completion date in your contract is 1st September?
L Yes, but it was extended to 1st October.
P When did you finish it?
L 1st December.
P Two months over your time.
L Yes.
P And liquidated damages for delay are £500.00 per month under clause 24.
L Yes.
P And the certificate you got when the moiety of retention money was released was endorsed 'Subject to clause 24 of the Conditions of Contract'. Why should liquidated damages of £1,000.00 not be deducted?
L Well, it's not fair. We did a good job, and everybody runs over contract time. It's recognised.
P Not under the contract. You agreed that if you did not finish on time you would pay damages.
L I never expected to have to. Nobody does.
P You got the extensions of time the architect thought you were entitled to get.
L We got some extensions.
P But you did not apply for any further extensions?
L No.
P And finished two months late?
L Well, it's not fair.

(Jorrocks – Angel, In-Chief)

J Are you Mr Herald Angel? Are you a quantity surveyor? Are you a partner in Messrs Angel, Gabriel and Partners of Great George Street, London, and have you been in practice as a quantity surveyor in this Contract?
A I am. I am. I am and I have.
J Who was responsible for the preparation of the bills of quantities in this Contract?
A I was.
J Whilst you were preparing those bills of quantities, did you inspect four trial holes on the site?
A I did.
J Naturally, what did you discover?
A Soil, sand and clay.
J Any rock?
A No.
J If you had suspected the presence of rock what would you have done?
A Measured it.
J How? Extra over any material?
A I could have done.
J Or separately?
A That is an alternative.
J And did you measure the rock?
A Yes.
J How?
A Extra over excavation in any material.
J Did you value that rock?
A Yes.
J How?
A In accordance with clause 13 of the Conditions of Contract.
J What was the measurement of that rock?
A 500 cubic metres.
J And the total value?

A	£12,000.00
J	Did you later prepare the Final Account?
A	Yes.
J	What sum did you include in the Final Account for the rock excavation?
A	£12,000.00
J	Did Mr Lucifer agree that sum?
A	The task of measuring and valuing variations under clause 13 is my duty, and in those matters my decision is conclusive. However Mr Lucifer did agree the sum of £12,000.00
J	Was your attention later drawn to arithmetical errors amounting to a deduction of £62 from your Final Account?
A	Yes. They were the result of the Borough Treasurer splitting hairs.
J	Did you consider that the sum of £62 should be deducted from your Final Account?
A	Yes, if one wishes to split hairs.
J	And you do?
A	No. Matters of principle are involved, and errors of fact or arithmetic discovered before the Final Certificate is issued should be corrected. The matter is of little importance.

(Pratt – Angel, Cross-examination)

P	Mr Angel. Your attention was drawn to the errors which had occurred by Mr Pluto on the 4th June.
A	They were not representations. They were a statement.
P	In fact, the Borough Treasurer did discover errors which you had made.
A	Yes, to the extent of £62.
P	And to the extent of £12,000.00 for rock.
A	That is not an error. Had a representation been made to me in respect of that matter I should have rejected it, as I informed the Borough Architect.
P	And to the extent of £1,000.00 for liquidated damages?
A	That is not an error, because it does not have to be dealt with within the Final Account.
P	Well, Mr Angel, you differ from me and I trust, from the arbitrator. However, I think we agree that the mathematical errors should be deducted?
A	Yes, £62.
P	So be it.

Pratt Opening

May it please you Sir,

The facts in this case have been given by my friend and so far as they are facts I am in agreement.

As far as my friend has stated, the real issues are as to whether the claimants should be paid for certain excavation in rock which occurred as the result of a Variation Order, whether or not the respondents are entitled to recover liquidated damages for certain delays which occurred, and as to whether there was wrongful interference by the respondents with the architect of such nature as to entitle the claimant to recover damages.

CALL MR PLUTO

(Pratt – Pluto, In-Chief)

P	Are you Mr Hector Pluto, the Borough Architect to the Borough of Barchester, the respondents, an Associate of the Royal Institute of British Architects, of the Guildhall, Barchester?
Pl	Yes, I am.
P	Are you acquainted with the Barchester Secondary School, the subject of the present dispute?
Pl	Well, I am, but not intimately.
P	You know Mr Angel?
Pl	Yes.
P	Did you receive the Final Account from his firm in March.
Pl	Yes.
P	Angel, Gabriel and Partners are the quantity surveyors for the school?
Pl	Yes.
P	And one of their duties is to prepare the Final Account?
Pl	Yes.
P	What did you do when you received the Final Account?
Pl	I glanced through it, but I know very little about the Contract. I passed it on to the Chief Quantity Surveyor in my department, and he examined it rather more fully.
P	Did he give you any report?
Pl	Not really. He handed it back to me, said he had looked through it but as it was Angel, Gabriel's account it would be all right.

P   What did you then do with the Final Account?
Pl I was going to issue the Final Certificate but my clerk said that by standing orders of the council all accounts had to be passed to Mr Midas.
P   Who is Mr Midas?
Pl The Borough Treasurer. I passed the Final Account to him.
P   What then happened to it?
Pl On the 24th May he passed it back to me with the memorandum which has been put in.
P   What did the memorandum state?
Pl That there were arithmetical errors amounting to £62, that payment for rock was not admissible and that there should be a deduction for liquidated damages.
P   Amounting to what sum?
Pl £13,062.00
P   £62 for arithmetical errors, £12,000.00 for rock and £1,000.00 for liquidated damages?
Pl Yes.
P   Did you form any conclusions?
Pl I thought there might be something in the Treasurer's argument about rock. The bills of quantities required the contractor to examine the site and trial holes, include for removing stone or rock and excavate in any material. It seemed to me that it might cover excavation in rock.
P   Is rock a material?
Pl So far as I know.
P   So the words 'any material' would include rock?
Pl I don't see why not.
P   What conclusions did you form about the liquidated damages?
Pl The contractor had made several written applications for extensions of time and there had been extensions to 1st October.
P   And the period during which the works remained incomplete under clause 24?
Pl Two months.
P   And the amount of liquidated damages in the Appendix to clause 24?
Pl £500.00 per month.
P   Now, Mr Pluto, will you examine this letter. Did you write this?

Pl Yes.
P   What does it say? Will you read it out?
Pl It states: 'In my opinion the works in the above ought reasonably to have been completed by 1st October'.
P   You signed the letter?
Pl Yes.
P   That is your certificate and the certificate required under clause 24?
Pl Yes.
P   And on that certificate the employer could deduct the liquidated damages due?
Pl Yes.
P   You did not issue a Final Certificate under clause 30.8?
Pl No. The Education Committee passed a resolution instructing me not to issue one.
P   And, of course, they were your employers and you accepted their instructions.
Pl I had to.

(Jorrocks – Pluto, Cross-examination)

J   Had you ever been named as architect in an RIBA Form of Contract before taking up your present appointment?
Pl No.
J   So you had never previously signed a final certificate?
Pl No.
J   Nor a certificate under clause 22?
Pl No.
J   Did Mr Angel measure rock in the manner referred to in the Standard Method of Measurement, as the contract required him to do?
Pl So far as I am aware.
J   Mr Angel was the person named to determine the matter in clause 13?
Pl Yes.
J   Not the Borough Treasurer?
Pl No.
J   Now let us return to the certificate under clause 24. Did anyone ask you to issue that certificate?
Pl Yes, the Treasurer, and I said I did not know enough about the circumstances to issue it.
J   But you did issue it.
Pl Well, the Town Clerk asked me to do so.

J Now let's turn to the Final Certificate. At the end of the Defects Liability Period, had all defects been made good to your satisfaction?

Pl Yes.

J Then why did you not issue the Final Certificate which that clause says you shall issue?

Pl Well, the Treasurer disputed the Final Account, and there was the Education Committee's resolution.

J But the surveyor, and not the Treasurer, was responsible for the priced bills of variations under clause 13?

Pl Yes, that is so.

J Mr Pluto, you are the person who must make the decisions and issue the Final Certificate, not the Treasurer.

Pl There was still the Committee resolution.

J Would you have accepted a resolution from the contractor as to what payment should be made?

Pl No, I suppose not.

J Then why should you accept one from the other party to the contract that payment should not be made.

(Pluto does not answer.)

J The real fact is, Mr Pluto, that differences existed between the Treasurer and the quantity surveyor and it was your duty to decide who was right under the contract. Is that not so?

Pl Yes.

J There was no difficulty about the arithmetical errors of £62. Both the surveyor and the Treasurer were in agreement that they should be deducted from that amount.

Pl Yes.

J And if any deductions were to be made for liquidated damages, it did not fall to be made by you, but by the employer's rights in that matter. You were in no difficulty there.

Pl If you put it that way, no.

J So that the only matter left was the rock, and that was a matter which by the provisions of clause 13 was left to the surveyor to decide?

Pl Yes.

J Could you not exercise your function of issuing the Final Certificate?

Pl I might have done.

J But did your employer interfere with the exercise of that function by forbidding you to issue the Final Certificate?

Pl I have no power to issue a Final Certificate after a dispute has arisen.

J But you were already out of time with your Final Certificate, and you had no notice in writing from either party under clause 30 of the existence of any dispute or difference.

Pl No.

J Thank you.

(Pratt – Pluto, Re-examination)

P Mr Pluto, whether you had notice in writing or not, a dispute or difference existed and you knew of its existence?

Pl Yes.

CLOSING ADDRESSES

(COMMENT – *It is not unusual for counsels' closing addresses, particularly those for cases of substance, to be made at a later date. When this is done the counsels may send summaries, in advance, to the arbitrator and use them as basis for their addresses. For the purpose of this case study, the closing addresses are given immediately after the examinations)*

ADDRESS – PRATT

Sir:
The issues in this dispute are clear.

With regard to the arithmetical errors amounting to £62, I gather that my friend does not propose to press his claim.

With regard to the excavation in rock, the bills of quantities provided that the contractor should inspect the site and satisfy himself as to local conditions, the full extent and nature of the operations and the execution of the contract generally, and further provided that the contractor should include in his tender for removing any naturally occurring stone or rock that might be encountered.

The contractor had therefore a duty to discover whether any rock existed upon the site, and if it did exist to include within his tender for its removal, but if that was not enough the bills of quantities went on to require the contractor to excavate in any material for the rates stated in the bills of quantities, and rock is a material, and having been paid for excavation in any material, it matters not that part of that material was rock. The employer has paid the contractor for excavating in any material and should not be required to pay him any further sums for the fact that that material was of a particular character.

With regard to the deduction of liquidated damages, there can surely be little dispute upon this point. The contractor had from time to time applied in writing for extensions of his contract period under clause 25, and those extensions have been granted to him. They had the effect of extending the contract period under clause 25 by two months, and the respondents do not dispute that he is entitled to benefit of that extension. Had there been proper grounds for further extensions he could have applied for such further extensions, but he did not do so. He cannot plead that he was ignorant of his right to do so, since he had already exercised that right. He failed to exercise it to that further degree. Extensions have been granted, and the period beyond the date of such extensions and the date of Practical Completion must rank for liquidated damages which the respondents should recover.

My friend will probably argue that the certificate which is required under clause 24 was issued as the result of pressure brought to bear by the respondents upon the architect. It was suggested to the architect by the borough treasurer and by the town clerk that he ought properly to issue that certificate, and in making such suggestions certain arguments were made as to why the certificate should be issued.

Whether the architect took cognisance of such arguments is beyond the point. He issued his certificate and he stated under cross-examination 'It was my decision and my certificate'.

That surely is sufficient to dispose of any suggestion that the certificate was wrongly issued as the result of pressure brought to bear upon the architect by the respondents.

The last question which arises is as to whether there was interference by the respondents with the architect in respect of his duties in issuing the Final Certificate.

It is true that the architect in issuing his Final Certificate acts in judicial capacity, and that under the circumstances stated in the contract he is bound to issue that certificate, but in order that an architect's certificate should be conclusive and binding certain conditions must exist, and one of those conditions is that that certificate shall be issued before a dispute has arisen. *Lloyd v Milward* (1895) and *Clements v Clark* (1880) are two of the decisions bearing upon that point.

(COMMENT – *If counsel rely on case law they will ensure that the appropriate reports are available for the arbitrator's consideration. The relevant cases should have been stated in the pleadings.*)

He has stated in evidence that he did not issue that certificate because of the resolution which had been passed by the respondent's Education Committee. But that resolution was only made because a dispute had arisen and the resolution was the result of the dispute. The respondents did not improperly prevent the architect from issuing a certificate which he himself considered he should issue.

This is not a case where in the absence of a dispute the respondents have said improperly to the architect: 'You shall not issue this certificate; you shall deprive the contractor of monies to which he is entitled'.

Indeed, the position is precisely the opposite. The respondents have said: 'We do not know what is the sum the contractor is entitled to receive, and until we discover that, a Final Certificate cannot be issued'.

How could they be in any way acting improperly?

The House of Lords in *Sutcliffe v Thackrah and Others* (1974) has affected the architect's immunity. He is now liable for negligence for every aspect of his duties, including when issuing certificates.

It has been suggested that the borough treasurer had no power to raise the issues which were raised by him but the respondents are responsible for the proper expenditure of public monies, and the borough treasurer is the officer appointed by them to advise them on the proper expenditure of those monies. He has done nothing more than the duties of his office compel him to do.

I therefore invite you to find that there is no further sum due to the claimants.

ADDRESS – JORROCKS

Sir:

In my Opening and that of my friend we agreed that substantially there were three matters with which you are invited to deal. Those three matters are as to whether the employer has any right to deduct any sum in respect of liquidated damages and as to whether we should be recompensated for the failure of the architect to issue his Final Certificate by reason of the interference of the employer, and the Quantum in each case.

We need spend no time on the arithmetical errors amounting to £62. We concede them.

With regard to the rock, a Variation Order was issued and the quantity surveyor determined that we were entitled to be paid extra for excavation in rock. He is the person appointed under the contract to determine that, and in the case of *Richards v May* (1883) it was held that where a contract provided that all extras and additions for which a builder should be entitled to payment should be paid for at the price fixed by a surveyor appointed by the building owner.

The quantity surveyor has exercised his functions in the manner which the contract provides. He has stated that he paid for the rock as an extra because the contract provided that the works should be measured in accordance with the Standard Method of Measurement, and that document provides that rock shall be paid for either as separately or extra over the various classes of excavation.

He is supported in that view by the case of *Bryant v Birmingham Hospital Saturday Fund* (1938). In that case, which was on a par with the present one, it was held that where the contract provides that work should be measured in accordance with the Standard Method of Measurement it should be so measured. It is not sufficient to include vague terms referring to the removal of stone or rock, or excavation in any material and the quantity surveyor did not for a moment suggest that it was sufficient.

With regard to the employer's right to deduct liquidated damages, it was a condition precedent to that right that the architect should give a certificate under clause 24 and he has stated that although such a certificate has been given it was the result of pressure by the employer.

It was wrong of the employer to exercise such pressure and the certificate ought not to have been issued. In the absence of such a certificate the employer would have had difficulty in recovering liquidated damages unless you, Sir, determined that he should.

My friend referred to the House of Lords decision in *Sutcliffe v Thackrah* and Others, but it must not be supposed that in giving this decision the House of Lords have in any way detracted from the position given to the architect under the Standard form of building contract by contractual agreement between the client and contractor.

Under these conditions the architect has certain duties clearly assigned to him. Thus he must, under clause 25 decide whether an extension of time is justifiable within the terms of that clause. His employer has agreed that this is the architect's function. The employer could not, without breach of his contract with the contractor, instruct his architect not to give an extension when the architect considers that under clause 25 the contractor is so entitled. Again, on the vital question of payment in interim certificates, the terms of the contract between the client and the contractor give the architect certain very clear duties.

Thus under clause 30 the employer had agreed that his architect shall issue interim certificates stating the amount due to the contractor from the employer and provided the architect is satisfied that the work is properly executed he must certify for the proper value of that work. If the employer, because it was not financially convenient to pay that month, were to tell him not to certify, or to certify a much smaller amount, the contractor would have clear grounds for claiming breach of contract; and the architect would be most ill-advised to accede to any such request by his client and indeed, of course, no professional person would wish to accede to such a request.

Subject to appeal under Article 5 the decision made by either the architect or the quantity surveyor in respect of any matter left under the contract to his discretion is conclusive, and the parties are bound by it.

My friend has argued that the architect has lost his powers under the contract by reason of some action taken by the employer. In this case a resolution made by a committee and confirmed by the council. That, of course, is not the case.

As to whether a dispute exists, however, before notice has been given under Article 5, that is a different matter, but one with which I need not worry you, since it is quite clear on the decision in *Brodie v Cardiff Corporation* (1919) and others that where an arbitrator having a jurisdiction has to decide that something ought to have been done by the architect which was not done, then if the terms of reference are wide enough to enable him to deal with the matter, as they are in this case, he may himself supply the deficiency and do that which ought to have been done.

You, Sir, may supply that deficiency by taking the place of the architect in respect of the Final Certificate, and in doing so you can restore the contractor, by means of damages or otherwise, to the position in which he would have found himself had that been done which ought to have been done.

I invite you, therefore, to say that the claimants should be paid that sum which the surveyor included in his Final Account in respect of rock, amounting to £12,000.00 that there should be no deduction in respect of liquidated damages, and that you should make good by way of damages the loss which the claimants have suffered by reason of the interference by the respondents with the architect, and so obstructing him from issuing his Final Certificate at the time it should have been issued.

(COMMENT – *The arbitrator may give directions regarding inspection, publication of his award and other relevant matters and close the hearing.*)

Before reading the award which follows re-read the synopsis and hearing and draft an award.

THE AWARD

IN THE MATTER OF THE ARBITRATION ACTS 1950 AND 1979
AND IN THE MATTER OF THE ARBITRATION BETWEEN:
JOHN LUCIFER AND SONS LIMITED (Claimant)
AND
THE COUNTY BOROUGH OF BARCHESTER (Respondent)

WHEREAS by an agreement in writing dated the first day of April between John Lucifer and Sons Limited (hereinafter referred to as the Claimant) of the one part and The County Borough of Barchester (hereinafter referred to as the Respondent) of the other part it was agreed that in case of any dispute or difference arising between the Claimant and Respondent as to certain matters set out in the said agreement such dispute or difference should be referred to the Arbitration and final decision of a person appointed on the request of either party by the President or Vice President for the time being of the Royal Institute of British Architects.

AND WHEREAS a dispute or difference having arisen between the parties to the said agreement the parties made joint application in writing dated the twenty-ninth day of July to the President of the Royal Institute of British Architects for the appointment of an Arbitrator and in the said application further did jointly and severally agree to pay the fees and expenses of the Arbitrator in accordance with any agreement in writing between him and the Arbitrator whether the Arbitration reached hearing or not.

AND WHEREAS the Vice President of the Royal Institute of British Architects did in writing on the thirtieth day of August appoint me NORMAN ROYCE of 3 Field Court, Gray's Inn, WC1 to be Arbitrator.

NOW I the said NORMAN ROYCE having accepted the said appointment and taken upon myself the burden of this Reference and having inspected the works and having heard and considered the allegations and witnesses and the evidence presented and the arguments and addresses made to me by Counsel on behalf of the parties, and having inspected the site.

DO HEREBY MAKE AND PUBLISH THIS MY AWARD in writing, touching and concerning the said dispute as follows:

1. that the Respondent shall pay to the Claimant the sum of £11,938 (exclusive of VAT) in full and final settlement of the claim;
2. that the Respondent shall pay to the Claimant the sum of £2,500.00 (exclusive of VAT) in full and final settlement of damages;
3. that the Respondent shall pay the Claimant's costs and the cost of the this arbitration upon a party and party basis;
4. that the Respondent shall pay my fees, costs and expenses of and incidental to the Hearing of the Arbitration and the drawing, making and publishing of this my Award which I tax and settle at £1,250.00 (exclusive of VAT);
5. that final settlement of the Award shall be effected within twenty-one days of the date of the Award.

(FIT FOR COUNSEL)

Made and published by me this twenty-ninth day of October.

_____
                                Arbitrator

In the presence of

_____
         Witness

_____
_____
_____
         Address

_____
         Occupation

4.12 Exercises

1. *Explain* the advantages and disadvantages of the alternative approaches to dispute resolution which have become increasingly popular in recent years and suggest reasons for their popularity.

2. *Discuss* the changes in the Arbitration Act 1950 introduced by subsequent Arbitration Acts 1975 and 1979.

3. *List and explain* the advantages and disadvantages of the arbitrator making a reasoned award.

4. A dispute has arisen on a building contract.
   Discuss the procedure to be adopted where:
   (a) the conditions of contract are JCT 80
   (b) one party has commenced legal proceedings
   (c) the conditions of contract have no arbitration clause.

5. Arbitration is normally preferable to litigation as a means of considering and settling disputes arising from building works.
   Examine this statement.

6. *Discuss* the following problems which could arise when a dispute has been referred to arbitration:
   (a) One party refuses to attend the hearing
   (b) One of the parties attending has briefed counsel
   (c) One of the contract documents has been destroyed by fire
   (d) A witness fails to attend the hearing.

7. *Prepare* an 'award' for the hearing in section 4.11.

8. *Identify* the circumstances in which an arbitrator's conduct may cause an award to be set aside by the courts.

9. *Discuss* the reasons for a disputant taking an 'offer to settle' seriously and explain why it might be costly if he does not do so.

10. *Discuss* the extent and nature of the influence which the courts are able to exert over the process of arbitration.
    (a) Outline the alternative procedures contained in the JCT Arbitration Rules 1988, rules 5 to 7
    (b) Highlight the differences between the procedures
    (c) Suggest reasons for the publication of the JCT Rules.

4.13 Sources and further reading

BADEN HELLARD, R, *Managing construction conflict* (Longman, 1988) provides a background to the evolution of English Law and the resolution of disputes including arbitration.

STEPHENSON, D A, *Arbitration for contractors* (Construction News Books, 1987) provides wider coverage of the arbitral process and contains a comprehensive collection of proforma, model orders, awards, etc. There is a section on the contractor as claimant.

GIBSON-JARVIE, R and HAWKER, G, *Guide to Commercial Arbitration under the 1979 Act* published by CIArb. In addition to being a useful guide it contains as appendices the Arbitration Act 1979 and useful extracts from the 1950 Act.

POWELL-SMITH, V and SIMS, J, *Construction Arbitration – a practical guide*, (Legal Studies and Services, 1989). The title indicates the contents. A text for the practitioner training to be an arbitrator.

JONES, N F, *A user's guide to the JCT Arbitration Rules*, (Blackwell Scientific Publications, 1989) is a lawyer's guide to the rules and includes specimen letters, etc.

Reference works include
Arbitration Acts 1950, 1975 and 1979, HMSO
Administration of Justice Act 1982, HMSO
JCT Arbitration Rules 1988, RIBA Publications
WALTON, A and VITORIA, M, *Russell on Arbitration* – 20th edition, Stevens 1982
BERNSTEIN, R, *Handbook of Arbitration Practice*, Sweet and Maxwell 1987
LEE, E, *Encyclopedia of Arbitration Law*, Lloyds of London, 1984

4.14 References

[1] ROYCE, N, Conciliation in business disputes: has it a future? *Construction Law Journal*, 1989 Vol 5 No.1, Sweet and Maxwell

[2] *ibid*

[3] GIBSON-JARVIE, R, and HAWKER, G, *A guide to commercial arbitration under the 1979 Act*, Chartered Institute of Arbitrators, 1980, chapter 1 explains the changes introduced by the 1979 Act in more detail.

[4] This section is taken almost verbatim from HAROLD CROWTER's paper in *Arbitration*, November 1989 issue.

[5] The author is grateful to Norman Royce for permission to reproduce the documentation and script for this practice arbitration hearing and award.

[6] A summary of the guidelines established by the Nema case is contained in JOHN SIMS' article, 'On the good ship Nema', *Building*, 15 January 1982. For full report see Weekly Law Reports (1981) 3 WLR, pp 295–305

[7] *Hansard*, Vol 398 (15 February, 1979) at col.1477.

Section 5
ALTERNATIVE CONTRACTUAL ARRANGEMENTS

5.1 Background to change

Until the 1960s a client with a need for building works would usually commission an architect to prepare drawings identifying his requirements. These drawings would provide the basis for competitive tenders by builders for the execution of the works. It is a system which was established early in the nineteenth century and which has continued for more than a century and a half. It is customarily referred to as the 'traditional system', or just 'traditional'.

The Emmerson and Banwell Reports In seeking an answer to why alternatives to the traditional system have evolved one might start with Sir Harold Emmerson who was asked by the Minister of Works in 1962 to make a quick review of the problems facing the construction industry.[1] His report included the now famous phrase that:

'in no other important industry is the responsibility for design so far removed from the responsibility for production'.

He concluded that the client suffered as a result of this 'divorce'.

Emmerson's Report led to the formation of the Banwell Committee[2] a year or so later which recommended a number of changes in contract procedures. The changes were, in themselves, significant but the most important effect of the Banwell Report was the change it engendered in the attitudes of central and local government. At that time some 60% of the construction industry's work was commissioned by central and local government. They were the industry's major clients and were in a strong position to dictate the contractual arrangements to be adopted.

Furthermore, the existence of a government commissioned report which encouraged government departments and local authorities to consider alternative approaches to building procurement made them less liable to charges of misconduct, failure to obtain the lowest tender, etc, etc. The relevant departments were able to take a wider view of public accountability; a view which was concerned not just with which tender submitted in competition was the lowest but which contractual arrangements facilitated the optimum overall result.

Most of the new arrangements claimed to facilitate shorter project periods, making earlier occupation possible and allowing the client to obtain an earlier return on his investment. The winds of change had started to blow through the construction industry!

The key issues identified by Banwell were:

- those who spend money on construction work seldom give enough attention at the start to defining their own requirements and preparing a programme of events for meeting them. Insufficient regard is paid to the importance of time and its proper use
- as the complexity of construction work increases, the need to form a design team

at the outset, with all those participating in the design as full members, becomes vital
- design and construction are no longer two separate fields and there are occasions when the main contractor should join the team at an early stage. The relationship between those responsible for design and those who actually build must be improved through common education
- some measure of selective tendering is preferable to 'open' tendering: impediments should be removed and rules for the conduct of selective tendering drawn up for the guidance of local authorities
- the use of unorthodox methods of appointing the contractor, where appropriate, has advantages which should not be lost to members of the public sector through adherence to outmoded procedures
- serial tenders offer great possibilities for continuity of employment; the development of experienced production teams, etc and the banding together of those who have suitable work in prospect is to be encouraged
- negotiated contracts need not be rigidly excluded in the public field; methods of contracting should be examined for the value of the solutions they offer to problems rather than for their orthodoxy.

Not withstanding the stimulus that the Banwell Report gave to change there were two other over-riding factors:

- the failure of the construction industry to satisfy the client's needs, particularly in respect of its management of exceptionally large and complex projects
- high inflation, coupled with high borrowing rates, which led to shorter project periods becoming of great importance to clients, particularly those who required an early return on their investment in property if the project was to be viable.

During the period 1973–74 many of the oil-producing states combined to bring about massive increases in the price of crude oil. The outcome was immediate, and there were massive increases in the borrowing rate and in inflation. The economy of the Western World was in disarray. A disarray which continued for more than a decade.

The increase in the public's and the construction industry's interest in alternative ways of procuring buildings more quickly was most marked following the oil crisis, as the industry's clients and their advisers realised that for many projects time was now of the essence.

There is little doubt that there was a positive relationship between the increase in borrowing rates subsequent to the oil crisis and the construction industry's interest in alternative systems of building procurement.

The cost effect of undertaking the design stage in parallel with the construction stage is demonstrated in the case study in section 5.3. Parallel working can produce significant reductions in the total cost of a project. These reductions are particularly pronounced when interest rates are high and when obtaining a return on investment made in the project is an important feature.

The growth of alternative systems for the procurement of buildings has had a significant effect on the role of the builder. His horizons are now wider. Previously, his activities were confined to carrying out the works. Now the builder is engaged as a management contractor, construction manager, as a member of a design-and-build team, or as a project manager and is able to work with members of the design team and advise the client on aspects of buildability which may have time and cost implications and so improve the overall viability of the project to the client's advantage. project to the client's advantage.

5.2 Procurement systems: types and terms

Figure 5.1 illustrates the alternative systems which have evolved. The four principal types are:

- designer-led, competitive tender
- designer-led, construction works managed for a fee
- package deal

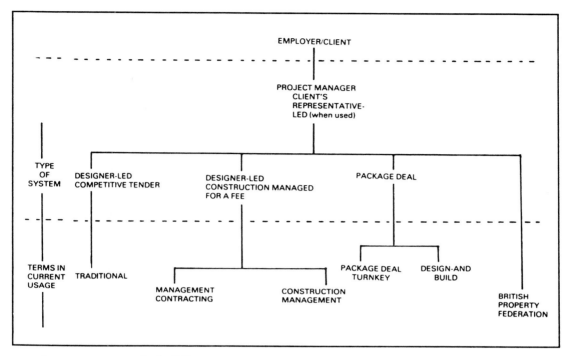

5.1 Alternative systems for building procurement

– project manager/client's representative led.

Various terms have emerged to identify the systems in current use, some of which are shown in figure 5.1. A glossary of terms in use is given in section 5.7.

The project manager/client's representative-led arrangement may be used in conjunction with any of the others.

Designer-led competitive tender Because most clients for construction work seek, at first, someone who can express their needs in the form of a design, the designer is, traditionally, the leader of the construction process. This 'traditional' approach provides a useful datum for consideration of the other systems available.

Traditional system The traditional system has evolved and developed over the centuries. The role of the architect was established in more or less its present form by the end of the eighteenth century by which time he was recognised as the independent designer of buildings and manager of the construction process.

Early in the nineteenth century bills of quantities began to be used as the means of providing a number of different contractors with a common basis for tendering. By the middle of the century the quantity surveyor was established as an independent compiler of bills of quantities and an expert in building accounts and cost matters.

There is considerable evidence, extending back over several centuries, of building craftsmen acting as contractors for complete building projects embracing the work of all crafts. Nevertheless, the general contractor in his present form is frequently regarded as coming into his own at the beginning of the nineteenth century.

The present traditional system which involves the parties mentioned above is enshrined in the Standard Form of Building Contract (with quantities).

Extent of use There are no reliable figures of the extent of use of the system but indications are that, perhaps, 60% to 70% of building projects, by value of the works, adopt the traditional system.

Operation The components of the traditional approach may be seen in a simplified form in figure 5.2. The process starts, as for all such processes, with a client having a need for a building (nodes (1) – (2)).

He briefs his architect on his needs, as he sees them, and by node (3) the cost ceiling for the project has been decided. The quantity surveyor should have provided preliminary cost advice by this stage.

Between nodes (3) and (4) the architect prepares alternative drawings/proposals so that the client may select that which he prefers; the quantity surveyor estimates the cost of the alternatives.

Between nodes (4) and (5) the client accepts a proposal.

Between nodes (5) and (6) the architect develops the design of the accepted proposal. This will probably entail consultations with specialist engineers and negotiations with specialist contractors.

Drawings and specifications are prepared and the quantity surveyor provides regular monitoring of the alternative designs to ensure that the cost implications of the design decisions are known to all concerned. The quantity surveyor prepares bills of quantities.

Between nodes (6) and (7), tender drawings, bills of quantities and forms of tender are sent to selected builders (contractors) in order that they may submit tenders for the work.

Beyond node (7) the builders estimate the costs of the operations involved in the project. The duration of the project is assessed from the pre-tender plan prepared by the builders' production planners and managers. Management decisions determine the margin to be added to the tender for profit.

In figure 5.2 the tender submitted by Builder A is accepted by the client and he and the builder enter into a contract (nodes (8) – (13)). The other tenders are rejected (nodes (9) – (12) and (10) – (11)).

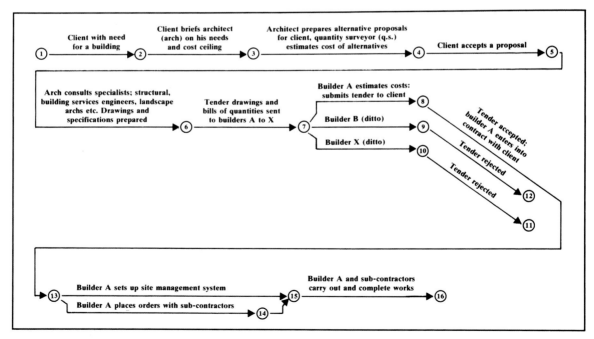

5.2 *The traditional system using standard form of building contract with quantities*

Between nodes (13) and (15) the contractor, Builder A, sets up his site management system, plans and organises the works, schedules materials' deliveries, etc. Concurrently, Builder A places orders with his own sub-contractors and those nominated by the architect (nodes (13) – (14)).

Characteristics of the system
- The system has operated in Britain, the Commonwealth and other parts of the world reasonably satisfactorily. It has stood the test of time.
- It is understood by most clients and they know their financial commitment when they accept the builder's tender if the design has been fully developed at the time of going to tender.
- The architect has considerable freedom to conceive and develop the design without excessive time or economic pressures, provided the cost ceiling is not exceeded and the client's requirements are generally satisfied.
- The project cost can be estimated, planned and monitored by the quantity surveyor from inception stage through to completion of the project.
- The system makes it possible for the architect to introduce consulting engineers, landscape architects and other experts to advise on or design 'sub-systems' of the project.
- The architect is able to consult specialist contractors and suppliers whom he believes to be appropriate for the project or who manufacture and/or install components for sub-systems which would be compatible with the system as a whole at design stage with a view to nominating them subsequently as sub-contractors or suppliers for the project.
- Sub-contractors may be invited to submit competitive tenders to the architect for the sub-system in which they specialise, thus ensuring that the most economic price is obtained.
- Drawings and bills of quantities provide a common basis for competitive tenders from selected main contractors.
- In the event of the client requiring the project to be varied during the course of construction, the bills of quantities contain prices for items of work which may be used to adjust the contract sum to take into account the variation(s).
- The design should be fully developed before bills of quantities and, subsequently, tenders are prepared. If not, excessive variations and disruption of the works are likely to occur.
- The need for the design to be fully developed before tenders are prepared leads to an 'end-on' design/build arrangement. Frequently, such an arrangement requires a longer overall project period than is necessary if both design and construction are able to proceed concurrently.
- As the length of the project period increases, so does the project cost because the client usually incurs financing charges on the sum which he has invested in land purchase, interim payments to the contractor and other members of the building team.
- Many contractors are of the opinion that their ability to organise and control the work of nominated sub-contractors is undermined by the nomination process, because such sub-contractors have less loyalty to the contractor than to the architect who nominated them.
- The separation of the design and construction processes tends to foster a 'them and us' attitude between the designers and contractors which reduces the team spirit that experience has shown to be vital for the satisfactory conclusion of a building project.
- Lines of communication between the parties tend to be tenuous and the interests of all may suffer as a consequence.
- The traditional system has been proved to be unsatisfactory for some large and complex projects which require advanced management systems, structures and skills.

Standard forms of contract   JCT 80 and IFC 84 may both be used with quantities and are appropriate for the traditional system.
   Standard forms of sub-contracts are used

for nominated sub-contracts under JCT 80 and for named sub-contracts under IFC 84.

The standard domestic (DOM) form of sub-contract is used with both JCT 80 and IFC 84.

Fast-track
The term 'fast-track' has been more subject to varying definitions than have other terms but overlapping of design and construction as a means of reducing project time is a generally recognised characteristic of the term. This overlapping, often referred to as 'parallel working', can be achieved by using a modified version of the traditional system or by adopting a form of construction management or management contracting.

Designer-led, construction works managed for a fee Under this heading are included the various management fee and construction management systems. There are almost as many variations on different systems as there are firms offering management services. The vast majority of the variations have one feature in common: the management contractor or construction manager offers to undertake the management of the works for a fee. He is, in effect, in much the same relationship with the client as is the architect or any other consultant. The actual construction work is undertaken by specialist contractors, each of whom contracts to carry out and complete one or more of the work packages which make up the whole of the works.

Those firms who adopt the title management contractor are normally those which are divisions of major construction contractor companies. One or two management contractors are now concerned solely with management contracting, having abandoned their original, traditional, contractor activities. The management contractor almost invariably employs the specialist contractors who undertake the work packages as his sub-contractors. It is the employment of the specialist contractors which typically distinguishes the management contractor from the construction manager. When a construction manager firm is employed the specialist contractors are generally in direct contract with the client, rather than being sub-contractors to the construction manager.

It is not possible to make categorical statements regarding who employs the specialist contractors (client or managing firm) because there are no codes of practice governing the operation of these contractual arrangements, but the broad generalisations made above may be taken as a reliable guide.

Two-stage tendering Two-stage tendering or two-tier tendering is another version of mangement contracting.

A statistically insignificant number of architectural firms adopt a successful variation of the construction management approach, the principal proponents being Moxley, Jenner and Partners. They refer to the approach as Alternative Method of Management (AMM). The client enters into separate contracts with the specialist contractors. The architect employs the site manager and undertakes responsibility for construction in addition to design.

The fee the contractor or construction manager, as the case may be, receives for undertaking management is not usually directly related to the value of the work being managed. It is not, for example, a percentage of the work value. In this way it cannot be said that the contractor has anything to gain if the value of work increases. The fee would, however, be renegotiated if the extent of the works changed significantly.

Management contracting/construction management The best known example of management/fee contracting in Britain is the Bovis system. Bovis have operated fee systems, notably with Marks and Spencer as a client, for more than 60 years with proven success. Very few other contractors operated such a system before the 1960s but management contracting is now well established as a procurement path.

Extent of use A significant percentage of major building projects, particularly those

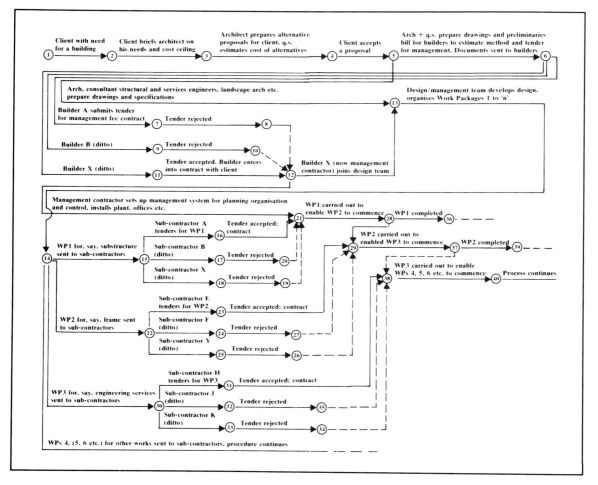

5.3 *Management (fee) contract system using two-tier tendering approach*

in the South East of England, adopt a 'fee' system of procurement. Nationally, however, probably no more than 10%, by value of the work, adopt a fee system. The use of this procurement approach appears to be declining but for major projects management for a fee systems have a considerable following of 'repeat' clients, regardless of statistical trends.

There is evidence that despite the overall decline in use, fee contracts are being adopted in some cities in the Midlands and the North where they had seldom if ever been used in the past.

Operation The process of a typical management contracting system is shown in figure 5.3. The construction manager's role is similar to that of the management contractor but he is less likely to be appointed by competitive tender. When a construction manager is appointed the works package contractors will most usually be in direct contract with the client.

For practical purposes 'construction manager' may be read for 'builder', 'contractor' and 'management contractor' in the following text under this heading, and in, figure 5.3.

It must be emphasised that the following 'operation' notes should be regarded as merely indicative of such systems.

Referring to figure 5.3 it can be seen that between nodes (1) and (5) the system is similar to that of the traditional system.

Between nodes (5) and (6), however, the architect and quantity surveyor concentrate on preparing drawings and a 'Preliminaries' bill of quantities in sufficient detail to enable the prospective fee contractor(s) to determine the method to be used for construction and to prepare a firm fee tender. At the same time the architect and other members of the design team develop the design generally and prepare drawings and specifications.

Beyond node (6) the contractor, (or contractors if the client seeks competitive tenders), prepares the first stage tender for the management fee. It is unusual for more than two or three contractors to be invited to tender.

By the time node (12) is reached Contractor X enters into a fee contract with the client and the other tenders are rejected.

The most competitive tender is often regarded as of less importance than a credible construction programme and a sound track record.

Extensive interviews with the staff of the contractors who are tendering are usually regarded by client and design team as an essential aspect of the selection process to ensure compatibility between design team and contractor's staff who will, if the tender is accepted, be working closely together as design-and-management team.

Between nodes (12) and (13) the management contractor, as he has now become, joins the design team. Concurrently, (nodes (12) to (21)), he establishes a management system for planning, organising and controlling the project. He installs the plant, site offices, etc.

Between nodes (13) and (14) the design-and-management team continues to develop the design and organises a series of work packages for all aspects of the work. The work packages provide the basis for a number of contracts which are placed as soon as the necessary information is available.

Beyond node (14) the work packages are put out to tender and contracts entered into. Drawings and bills of quantities or specifications may be used as the documentation for the sub-contract tenders. Having between 30 and 40 work packages is by no means unusual. For major projects the number of work packages may be as many as 150.

The works contained in the work packages are frequently commenced almost as soon as the contracts have been placed. Project completion is achieved with completion of all the work packages.

Characteristics of the system
– Management contracting has been used successfully to a limited extent since the 1920s and with increasing frequency during the 1970s and 1980s.
– Clients and contractors often adopt the system on a regular basis once they have gained experience, which suggests that it has merits. It is generally recognised that its adoption requires mutual trust.
– Work can commence as soon as design proposals have been accepted by the client and drawings have been approved by the local authority.
– The management contractor (or construction manager) is appointed much earlier than would be possible with the traditional system. He is able to become a member of the design team and contribute his construction knowledge and management expertise.
– Management contractors (or construction managers) frequently compete at first stage tender ensuring that an economical fee is charged for management.
– 'Them and us' attitudes are reduced and lines of communication are improved.
– The management contractor finds it easier to identify with the client's needs and interests and 'integration of the team' becomes possible and practical.
– Decisions regarding appointment of sub-contractors are made jointly (by designers and construction manager or

management contractor) thus making use of wider experience.
- Specialist contractors (or sub-contractors) compete at second stage tender ensuring economical tenders.
- Contracts are entered into near the time of commencement of the works making firm-price tenders possible.
- Tenders submitted near the time of commencement of work are frequently more competitive than those submitted several months or even years ahead.
- When a construction manager is employed, the client enters into contracts with numerous specialist contractors instead of with a general contractor as would be the case if the traditional system were adopted. He usually has a closer involvement in the project throughout its whole life.
- Lines of communication between clients and specialist contractors are shorter than with the traditional system. Advantages which stem from this factor are:
- the client is enabled to make prompt decisions which can be implemented without delay; it makes possible a prompt response by the client to unforeseen site problems and by the contractors to changes required by the client
- the cost implications of design changes can be promptly assessed and cost control for the client is therby facilitated.
- Specialist contractors frequently prefer to be in contract with the client rather than with a management contractor because interim payments are usually made more promptly when paid direct.
- When contracts are made direct between client and specialist contractor, conditions of contract can be adopted which are appropriate to the needs of the works to be undertaken.
- The total project completion period is reduced by parallel working.
- A reduced project completion period produces a corresponding reduction in financing charges on the sum invested in land purchase, interim payments to contractors and other members of the building team. Inflation has less effect.
- The client takes delivery of the building more quickly because the project completion period is reduced and thus obtains a return on his investment more quickly.
- The client is usually given an approximate estimate of the final project cost by the quantity surveyor and/or contractor early in the project life, but he does not know the final project cost until the last sub-contract is entered into. On other projects he is given a guaranteed maximum cost.
- The architect may have less time to develop the design because he is under greater pressure from client, contractor and sub-contractors. The design may suffer as a result.

Standard forms of contract The JCT Standard Form of Management Contract, 1987 edition (MC 87), may be used for contracts between the employer (client) and management contractor.

Works contracts WC/1 and WC/2 are used for contracts between the management contractor and the various works (sub-) contractors.

Package deal design-and-build Under this heading are included terms such as turnkey, package deal, contractor's design and design-and-build systems. To all intents and purposes the terms turnkey and package deal have the same meaning.

The range of services offered by package-deal contractors varies greatly. Some will find sites, arrange mortgages, sale-and-leaseback and similar facilities in addition to designing and building to meet the client's requirements. Others contract to design and build a unique building on the client's own site. The feature that the systems have in common is that the 'contractor' is responsible for the whole of the design and construction of the building. Responsibilities are not split between designer and builder so that the client finds himself looking to separate 'parties' in the event of a building failure. The systems offer 'single-point responsibility', a feature that commends itself to clients frustrated by the traditional system.

Package-deal contracts involve direct negotiation between client and contractor (or several contractors if the client seeks competitive tenders). The client states his requirements and the contractor (or contractors) prepares design and cost proposals to meet the requirements. Initially the contractor produces only sufficient by way of design proposals to demonstrate his 'package' to the client. The design is fully developed when both parties have reached an agreement regarding specification and price.

Experience indicates that clients are frequently able to procure buildings more quickly when these contractual arrangements are adopted. Time savings tend to go hand in hand with cost savings.

Design and build is a more refined form of package deal which obtained recognition from the Joint Contracts Tribunal in 1981 with the publication of the JCT Standard Form of Building Contract with Contractor's Design (CD 81). This recognition followed changes in British architects' codes of practice which allowed architects to become directors of construction firms. Hitherto they were able to be salaried employees but director status had been denied. Figure 5.4 shows the stages in the operation of CD 81.

There is evidence that package-deal design standards have improved as architects have taken up senior appointments in design-and-build firms or, as is by no means unusual, founded firms which are predominantly designer led.

Most types of building have been constructed using a package-deal approach but industrial and office buildings in new development areas are the most typical examples of building types which are frequently built using a package deal system. Many package deals involve a proprietary building system of one sort or another. Package dealers frequently advertise their services and/or product in the pages of newspapers and journals read by the people who are likely to make decisions regarding their firm's future building needs. Package deals provide buildings rather than designs and the dealer may offer to find a site in the part of the country where, for example, government grants are available to the client, in order that he has an incentive to expand his business in that area – high unemployment areas, for example.

The package dealer will usually undertake to obtain planning permission and building regulations approvals.

Extent of use It has been estimated that between 15% and 20% of building projects, by value of works, are carried out by some form of package deal. The use of this procurements approach appears to be increasing.

Operation For purposes of illustration the contractual arrangement and terms used in CD 81 have been used.

The components of the system may be seen in figure 3.4. It starts when the client identifies his need for a building between node (1) and (2), and states his requirements between nodes (3) and (4). In practice he may ask an 'agent' to prepare his 'Employer's Requirements', as they are termed in CD 81. The client might employ an architect, quantity surveyor, building surveyor or similar competent person to state his requirements but such a person's task would be complete when he had prepared the statement.

The client's requirements are passed to the design-and-build contractors, nodes (3) to (4), each of whom prepares his design and ascertains the time it will take him to carry out the works. At the same time he prepares an estimate of the cost of his 'proposals' and submits a tender (following node (4)). No more detail is given than is necessary for tender purposes.

The client's requirements need to be submitted only in sufficient detail to enable the contractors to ascertain needs and submit their proposals. In figure 5.4 three

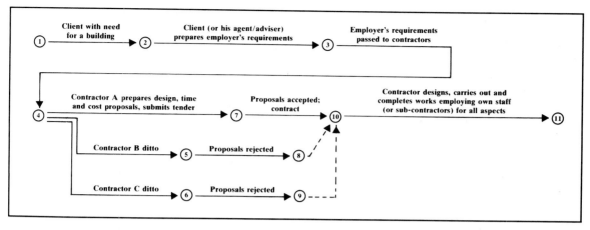

5.4 *Design-and-build system*

contractors are shown to be submitting proposals but it is by no means unusual for a client to negotiate with only one contractor.

In figure 5.4 the proposals of Contractors 'B' and 'C'; are rejected but those of Contractor 'A' are accepted (between nodes (5) and (10)). Contractor 'A' now prepares a detailed design and carries out and completes the works employing his own staff or sub-contractors.

The contractor's proposals normally include a 'Contract Sum Analysis', which takes the place of bills of quantities. It is generally accepted that the Contract Sum Analysis should contain sufficient pricing data to enable the cost of 'changes in the Employer's Requirements' to be calculated, should changes occur.

There is provision in CD 81 for the client to nominate an 'employer's agent' whose role is to receive or issue applications, consents, instructions, notices, requests or statements or to otherwise act for the employer. This agent will probably, but not necessarily, be the man who prepared the statement of client's requirements. He has a much more restricted role than that enjoyed by the architect or supervising officer when the traditional system of contracting is used.

Characteristics of the system
- It is used increasingly as a means of managing the building process at home and abroad.
- It provides single point responsibility so that in the event of a building failure the contractor is solely responsible. There can be no question of 'passing the buck' between architect and builder as has so often been the case in the past. The client's interests are safeguarded in this respect.
- The client knows his total financial commitment early in the project's life, provided he does not introduce changes during the course of the works.
- The client has direct contact with the contractor. This improves lines of communication and enables the contractor to respond and adapt more promptly to the client's needs.
- The contractor is responsible for design, construction planning, organisation and control. These activities can proceed concurrently to a greater extent than is generally possible using the traditional system.
- The package dealer may provide a comprehensive package comprising site seeking and purchase, obtaining planning permission and building regula-

tions approval, financing facilities, leasing, etc.
- The package dealer may use a proprietary building system or modular building form which reduces design time and the time required for approval of the building components.
- The client is frequently able to see examples of the package-dealer's product when his proposals are being made. Most clients can visualise their needs more readily in three dimensions (by moving within and 'sampling' an actual building) than by the study of drawings and specifications. Quality, a feature which it is difficult to specify, may be more easily indicated by comparison with a sample.
- Many systems used by package dealers have been tested over a period of years and are less prone to teething troubles.
- There have been some serious failures among building systems.
- The package-dealer's components are often readily available so that manufacturing time is minimal and construction time may be correspondingly reduced because manufacture of components and work on site can proceed concurrently.
- Work on the building can commence as soon as local authority approvals have been obtained and sufficient information regarding the earlier site operations is available. The design does not need to be finalised before some, at least, of the work may be commenced.
- The package dealer is familiar with the construction methods to be used for his product and work proceeds more quickly.
- Some proprietary package-deal products lack aesthetic appeal.
- The range of designs available from some proprietary package dealers is sometimes limited.
- Competition between the contractors' proposals should ensure economical tenders and alternative design concepts.
- The relaxation of the architects' code of practice makes it possible for them to become full partners in design-and-build firms.
- This relaxation should lead to the construction of buildings which reflect the senior status of the designer in the team and lead to more aesthetically pleasing buildings than may have been built in some instances in the past.
- The nature of the system should promote the creation of an integrated design and construction team.
- The closer involvement of architects in the building process should lead to designs which have a greater appreciation of construction methods; 'buildability'.
- The integrated nature of the team improves communication between designer and builder which encourages prompt decisions.
- A prompt response in the event of materials or manpower shortages.
- Design costs are built into the package but because the design input and 'detailing' required are less than when using the traditional system the costs involved are frequently less.
- There is no independent architect or similar 'professional' available to the client to advise on the technical quality of the designs at time of tender, although he is not precluded from seeking such advice if he so wishes.
- The employer's agent may supervise the works and ensure that the contractor's proposals are complied with and that the work is not skimped.
- The nature of the contract tends to reduce changes (variations) from the original design and disruption of the works is less likely to occur.
- The reduction of changes and disruption produces time and cost savings which benefit the client.
- The total project completion period is reduced.
- Time savings reduce the employer's financing charges, inflation has less effect and the building is operational sooner which, in a commercial context, produces an earlier return on the capital invested.

Standard forms of contract CD 81 is intended for use on projects where the client provides the site which is the subject of the contract. Many design-and-build projects use conditions of contract drafted for specific purposes. The diverse nature of these projects leads to correspondingly diverse conditions of contract.

DOM forms of sub-contract may be used between contractor and the various sub-contractors.

Project manager/client's representative
During the 1960s and 1970s construction projects tended to become larger and more complex. It became apparent that the time-honoured client-architect-builder relationship was sometimes inadequate as a system for constructing buildings within cost-budgets and tight time-schedules. There was a need for someone to manage the project as a separate, distinct member of the construction team – a project manager or client's representative.

There is nothing new in the concept of a project manager. Before the end of the seventeenth century when architecture, as a profession, was established in Britain, virtually all major building projects for Church and Crown – the principal clients of the building industry – were designed by craftsmen and managed by influential 'clerks' who were frequently known as Clerk of Works or Master of Works.

These men were the client's representative. They held the purse strings and they had overall management of the project. The emergence of project managers for major projects in the 1960s marked the return to a system which existed for some 600 years in Britain.

The essence of the appointment of a project manager or client's representative is that a single person acts as surrogate client. The title project manager is that which is most generally employed but client's representative is becoming increasingly used. Whichever title is used the role is to ensure that all the needs of the client are satisfied and to act as the contact point between client and the building procurement team. It is the direct relationship between client (whose interest the project manager represents) and the project manager which distinguishes his role from other 'managers' in the construction process who frequently have the word 'project' affixed to their 'manager' title.

The Wood Report[3] suggests that the project manager's prime task is one of co-ordinating client requirements such that clear instructions from a single source can be provided to the other parties involved. The importance of the client identifying a single person to represent his interests (before he has a firm commitment to actually building) and to assist him with drafting the brief for the project is recognised in *Thinking about building*,[4] which provides a guide to the selection of the most appropriate procurement path to meet a potential client's specific needs.

A vital feature of the project manager's/ client's representative's role is that he is concerned solely with managing the project. Because he is not involved in designing or constructing the building works he is able to take an objective overview of the activities of all concerned.

In 1988 the NEDO report *Faster building for commerce*[5] identified the need for the client (referred to in the report as the 'customer') to appoint an experienced 'customer representative' with experience in working with the construction industry if his in-house project executive was insufficiently experienced. The report suggested that such a person can be found among architects, engineers, surveyors, project managers or in contracting companies with management and/or design skills as well as those of construction.

Such representative must have sufficient status and authority to act on the client's behalf in the dialogue between the client's organisation and the team appointed to procure the building. There is no reason to believe that the requirements of clients for commercial buildings differ greatly from those of clients for other types of building.

The relationship between the parties to the contract when a project manager or client's/customer's representative is employed is shown in figure 5.5.

Extent of use Project managers have been engaged with increasing frequency and

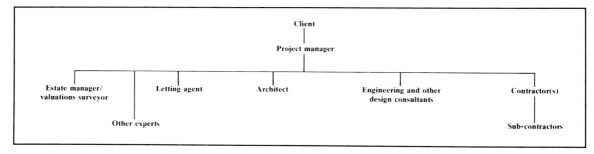

5.5 Relationship between parties for project management systems

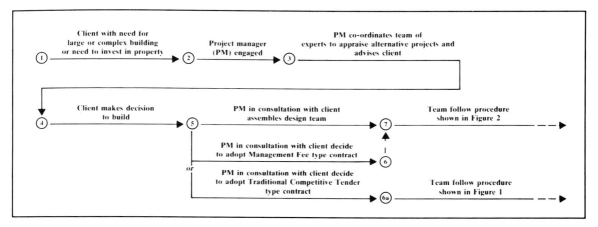

5.6 Project management

success during the 1970s and 1980s but there is no record of the extent of use.

Operation The components of the system may be seen in figure 5.6. As with the systems described previously there is a client with a need for a building. Between nodes (2) and (3) a project manager is engaged.

The first task of the project manager is to appraise alternative ways of meeting the client's needs; (nodes (3) to (4)). If, for example, the aim of the project is to maximise the client's return on his investment, the project manager might well consult letting agents or undertake extensive market research. On high technology projects, such as nuclear power stations, the appraisal might be between alternative energy sources. The magnitude of many projects requiring the engagement of a project manager is such that his role at this stage is 'co-ordinator of expertise'. He will then present the experts' collective recommendations to the client for his decision regarding the project which best suits his needs (between nodes (4) and (5)). In practice the client will more usually be the board of directors of a corporation or the council of a public authority.

Nodes (4) to (5) mark the point at which a decision has to be made regarding the selection of the contracting system to be employed. If some form of management fee system is to be employed the contractor may be appointed at this stage. In this event he will join the design team and the remarks made about this system will apply.

Between nodes (5) and (7) the project

manager assembles the design team which will best suit the project's special features, whilst at the same time endeavouring to appoint people who will work well together. An appreciation of the human aspects of management is an important part of the project manager's skills. Beyond node (6) or (6a) the project is designed, costed and constructed. It is important that the cost implications of design variables are ascertained as promptly as possible so that their effect on project viability can be considered by the project team and appropriate action taken.

If the contractor has not been selected at node (6), nor joined the design team at that point, he may be selected beyond node (6a) by competition, using the traditional system.

Alternatively, a package deal contractor may be appointed, in which event the project manager's role will be mainly concerned with acting on the client's behalf in the dialogue between his organisation and the contractor.

Characteristics of the system
— It has been used with increasing frequency and success during the 1970s and 1980s for complex and large projects.
— It is popular because of the dissatisfaction of some clients with the traditional system and its associated delays and excessive costs.
— The project manager is a 'professional', surrogate, client with experience of identifying and stating the client's needs and requirements.
— A project manager will often have a professional background appropriate to the type of building to be constructed.
— The project manager is able to act as a leader who can take into account all aspects of the project; finance, feasibility, design and time and hold a balance between them.
— The engagement of a project manager releases the client from the need to delegate a member of his staff (often a member without previous experience) to act as the intermediary between client and project team.
— The management function is separated so that the manager is able to act in an independent capacity.
— The client incurs an additional cost from the project manager's fee but this cost is offset to some extent by savings in his own 'management' involvement.
— The design and construction functions are separated so that those involved can act as partners on equal terms.
— 'Them and us' confrontation may be avoided as a result of this separation.
— Overall project planning and control which results from the engagement of the project manager ensure that both design and production are planned and co-ordinated to give as short an overall design and construction duration as possible.
— The architect and other consultants are released from the tasks and problems associated with managing a project, enabling them to concentrate on design matters.
— The quantity surveyor carries out cost estimating, and planning and control throughout the overall project period.
— The system is able to combine the advantage of the traditional system, which is understood by most clients, with improved management methods.
— The system provides for alternative means of selecting the contractor.
— Reduction of the overall project period provides consequential cost reductions as the client is able to utilise the building, or obtain a return on his investment, more promptly.

Standard forms of contract There is no JCT standard form of contract for the employment of a project manager. JCT 80, IFC 84, CD 81 or MC 87 may be used between client and contractor. The appropriate sub-contract forms may be used between contractor and sub-contractors.

The British Property Federation System
The British Property Federation (BPF) is a powerful client body which has recognised the importance of the client appointing a single person to represent his interests. The

BPF has done much to promote the term 'client's representative'.

In November 1983 the British Property Federation published a Manual of the BPF System for building design and construction. The manual comprises 99 pages of which 36 are appendices providing schedules of responsibilities, checklists and proformas.

The manual excited considerable interest and criticism because it proposed radical changes to established procedures. Some members of the building team saw their traditional roles threatened.

The BPF system 'unashamedly puts the client's interests first'. It attempts 'to devise a more efficient and co-operative method of organising the whole building process ... to the genuine advantage of everyone concerned in the total construction effort'. The reason for this enterprise is that 'to build in this country costs too much, takes too long and does not always produce credible results'.

The BPF represents substantial commercial property interests and thus it was able to exercise considerable influence on the building industry and its allied professions, particularly at a time when the industry was working at much less than its optimum capacity.

The BPF manual is the only document of its kind which sets out the operation of a system in such detail. Clearly, the manual provides the definitive document and the 'operation' described below should be regarded simply as an introduction to the system. This disclaimer is significant because the manual is at pains to offer a system which can be used with various methods of contracting and one which 'although consisting of a series of precisely described steps, can be used flexibly'. In many respects the system is an amalgam of those discussed above.

Extent of use The use of this procurement approach is increasing but measured by value of work, the system's contribution to the building industry's output is not great. Nevertheless, the system has made a significant contribution to developing the industry's attitudes and approaches to building procurement.

Operation The components of the system may be seen in simplified form in figure 5.7. The process commences at node (1), when a client plans to build.

BPF members are largely 'commercial' but the Federation's system should be capable of adoption by a wider range of clients.

The manual suggests that the client should explore the many courses open to him 'at minimal cost' and appoint a 'client's representative', who is defined as 'the person or firm responsible for managing the project on behalf of and in the interests of the client'. The client's representative may be an employee of the client or an architect, chartered surveyor, engineer or project manager.

At node (3) the client appoints the client's representative whose first tasks are to help the client develop the concept and manage the project on his behalf. Obviously, the extent to which it will be necessary for the client's representative to become involved in ascertaining the economic viability and technical feasibility of the project will depend on the client's in-house skills and expertise.

Between nodes (4) and (5) client and client's representative develop the outline brief and the outline cost plan to the point where the client is satisfied and to the extent that the full brief may be specified. The activities between nodes (1) and (4) comprise 'Stage 1 – Concept' in the BPF manual.

Node (4) sees the commencement of 'Stage 2 – Preparation of the brief'. It is possible that the 'design leader' may have been appointed in Stage 1 but if not, he will be appointed at node (5).

The design leader is defined as 'the person or firm with *overall responsibility* for the pre-tender design and for sanctioning the contractor's design'. The design leader might be an individual, a multi-disciplinary firm, or he might be a consultant with specialist consultants contracted to him. The words 'overall responsibility' have

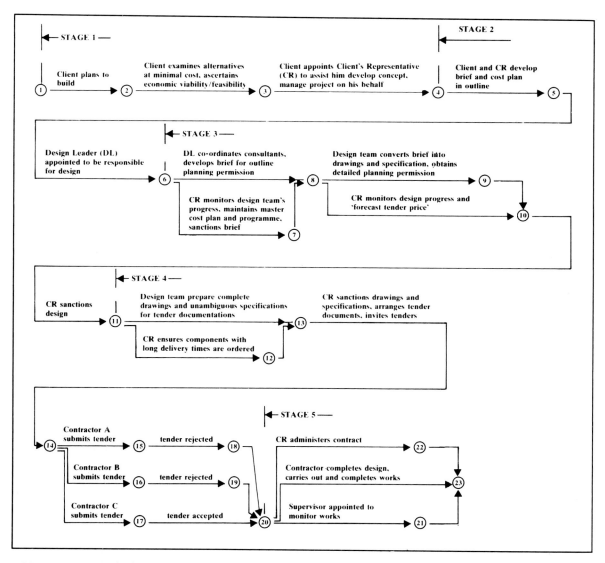

5.7 *BPF system for building design and construction*

been printed in italics above to emphasise the 'unique' role of the design leader.

'Stage 3 – Design Development', commences at node (6). Between nodes (6) and (7) the design leader co-ordinates consultants and develops the brief to the point where an application for outline planning permission may be obtained from the local authority.

Concurrently, the client's representative monitors the design team's progress and prepares and maintains the master cost plan and the master programme. The former is 'a schedule prepared by the client's representative of the expenditure required to implement the project' and the latter is 'a schedule prepared by the client's representative of the main activities to complete the project'.

By node (8) the client's representative

will 'sanction' the design leader's brief and, subject to obtaining outline planning permission, the design should progress to the point where an application for detailed planning permission may be obtained, (nodes (8) to (9)).

The glossary of terms refers to 'sanction' as 'the process by which the client's representative successively agrees the work of the design team to ensure that it meets the requirements of the brief'. The contractor's design is similarly sanctioned by the design leader to ensure that it complies with the contract documents.

The client's representative continues to monitor design progress and at node (10) agrees the 'forecast tender price' which is a 'forecast made by the design leader of the likely cost of construction'. The forecast tender price forms part of the master cost plan, referred to above.

Between nodes (10) and (11) the client's representative sanctions the design as far as it is advanced at this point.

Between nodes (11) and (12) and (11) and (13) design leader and client's representative work together towards the provision of tender documentation. The design team prepares, what are referred to in the manual as, 'complete drawings' but this term may be misleading if the reader is accustomed to the traditional system in which design is entirely the province of the design team. Complete drawings in the context of the BPF system mean that the drawings, together with 'clear unambiguous specifications', are sufficient as a basis on which contractors might tender without 'being justified in claiming for omissions or inadequate descriptions'. The manual points out that the quality of the information will control the standard of the buildings.

Between nodes (13) and (14) the client's representative sanctions the drawings and specification, arranges tender documents and invites tenders. 'Stage 4 – Tender documents and tendering' commences at node (11) and is completed at node (20) when a tender is accepted and a contract is entered into between client and contractor.

There will probably be a need for clarification of sundry items by all concerned with the project and the prospective contractor may be required to provide further information, costs, calculations, etc, before contracts are finally exchanged.

Tender documents consist of:

– invitation to tender with its appendices
– specifications
– drawings
– conditions of contract
– bills of quantities, should the client decide to use them.

The contractors' tenders are submitted to the client's representative.

The tender should include:

– outline priced schedule of activities
– organisation chart
– details of personnel
– method statement
– list of declared sub-contractors
– schedule of time charges.

The tender may also contain alternative proposals for the design and construction of the building.

'Stage 5 – Construction' is carried out between nodes (20) and (23). A particular feature of the BPF system is that the contractor 'completes the design, providing co-ordinated working drawings. He obtains approval of his design from statutory authorities should this be necessary and co-ordinates the work of statutory undertakers'. The building agreement between the client and the contractor states that the contractor's design is to be sanctioned by the design leader to ensure that it complies with the tender specification.

The client's representative administers the building contract, approves payments to the contractor, decides on the need for variations and issues instruction. It is he who decides if the services of the design leader should be retained during the construction stage (nodes (20) – (23)). It will be appreciated that the design team's tasks should have been completed by node (20); perhaps by node (14).

A supervisor is appointed to monitor the works (nodes (20) – (21)); his duties are detailed in the appendices. They are similar to those of a clerk of works but more comprehensive in their scope.

Characteristics of the system
- It was devised, almost unilaterally, by one party to the building contract – the client – so it lacks some of the compromises inherent in agreements devised by bodies such as the Joint Contracts Tribunal. It is concerned primarily with the client's interests.
- It is designed to produce good buildings more quickly and at lower costs than the traditional system.
- It is designed to change attitudes and alter the way in which members of the professions and the contractors deal with one another with a view to creating a fully motivated and co-operative building team and to removing as much as possible of the overlap of effort between designers, quantity surveyors and contractors which is prevalent under the traditional system.
- It is designed to redefine risks and re-establish awareness of real costs by all members of the design and construction team and to eliminate practices which absorb unnecessary effort and time and obstruct progress towards completion.
- It provides for an independent 'client's representative' who manages the project as a whole and who is not involved as a designer or contractor. He provides single-point responsibility for the client and by virtue of his non-involvement in details he is able to concentrate on management.
- It creates a design leader with overall responsibility for the pre-tender design and for sanctioning the contractor's design.
- The contractor's knowledge and experience of the cost implications and buildability of design variables may be utilised to good effect because he contributes to the design.
- It provides financial incentives which encourage contractors to undertake design detailing that is economical to construct.
- The arrangement by which the contractor undertakes detailed design should reduce 'pre-tender' time and so enable the client to have earlier occupation of the building and an earlier return on his investment. He should incur lower financing costs because of a reduction in the overall project period.
- The system makes provision for the design team and contractor to negotiate upon and alter the pre-tender design before entering into a binding contract. This should reduce variations once the works are in progress.
- There is provision for the design team to name sub-contractors and suppliers who they would require (or prefer) to be invited by the contractor to tender for part of the works. There is no provision for nominated sub-contractors as with the traditional system.
- It supports the use of specifications, rather than bills of quantities, as the basis for obtaining competitive tenders from contractors despite the preferences of contractors and others for bills of quantities.
- The contractor is required to provide, as a tender document, a priced schedule of activities which supplants bills of quantities and may be used for managing the construction works, monitoring progress, ascertaining the amounts of payments on account to the contractor and negotiating the value of variations.
- Consultants' fees for their services are subject to negotiation rather than being determined by closely defined 'scales' as has been the custom with the traditional system.
- An adjudicator is appointed to decide impartially disputes which may arise in the implementation of the project arrangements. His task is to carry out a prompt investigation and give a decision which is implemented forthwith. There is provision for reference to arbitration 'after the taking over of the works' if

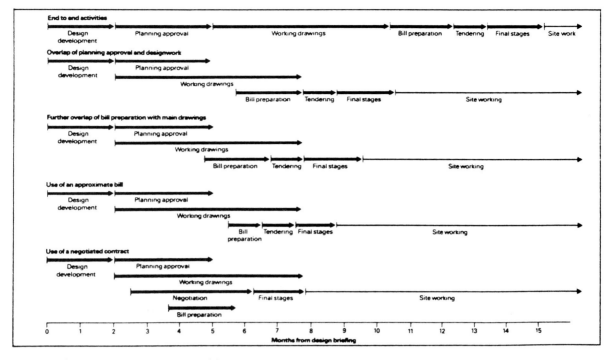

5.8 The pre-construction timetable
Faster building for industry, NEDO, HMSO, 1983

the dispute cannot be resolved by the adjudicator.

Form of contract The ACA form of building contract, BPF edition is available for use with the BPF system.

5.3 Project time: cost relationships

This publication contains numerous references to the time and cost advantages of various systems for the procurement of buildings.

This section aims to illustrate the extent of the relationship of time to cost.

The report *Faster building for industry* states that that the traditional methods of design and tendering, the various design-and-build options and project management can all give good construction times but on average, the use of non-traditional routes tends to produce overall times shorter than those produced by the traditional routes. Figure 5.8 provides a 'pre-construction timetable' which demonstrates the relative times of the options.

From figure 5.8 it may be seen that the use of a negotiated contract enables construction work to commence some seven months earlier than is possible with the 'end-to-end' traditional system. It is reasonable to assume that whichever approach is used for the design, the construction periods will not differ significantly so that the total project period will be reduced if one of the non-traditional or 'fast-track' approaches is used. What effect does this have on the project cost?

Case study illustrating the relationship between time and cost

The case study relates to a commercial project in a city centre.

PROJECT TIME: COST RELATIONSHIPS

	Cost £s K	Months 3	6	9	12	15	18	21	24
Site acquisition	6,000	6,000							
Design	1,200	250	250	250	250	200			
Construction	10,000					2,500	3,000	3,000	1,500
Quarterly total		6,250	250	250	250	2,700	3,000	3,000	1,500
Cumulative total		6,250	6,680	7,123	7,579	10,499	13,733	17,055	18,977
Financing cost (quarterly)		180	193	206	220	234	322	422	524
Total	17,200	6,430	6,873	7,329	7,799	10,733	14,055	17,477	19,501

(a) Expenditure plan — traditional system

	Cost £s K	Months 3	6	9	12	15	18	21	24
Site acquisition	6,000	6,000							
Design	1,200	300	300	300	150	150			
Construction	10,500		3,000	3,000	3,000	1,500			
Quarterly total		6,300	3,300	3,300	3,150	1,650			
Cumulative total		6,300	9,780	13,274	16,723	18,780			
Financing cost (quarterly)		180	194	299	407	514			
Rental income $2.3m pa)							(575)	(575)	(575)
Total	17,700	6,480	9,974	13,573	17,130	19,294	18,719	18,144	17,569

(b) Expenditure plan — fast-track approach

5.9 *Time:cost case study for commercial project*

Figure 5.9 shows alternative expenditure plans for the project assuming use of the traditional system and a 'fast-track' approach.

The data have been provided by a client body specialising in such projects.[6] Some of the data cannot be more than indicative but the study provides a reasonably reliable basis for cost comparison.

Figure 5.9 (a), the expenditure plan for the traditional system, shows an estimated design period prior to commencement of construction of 12 months. Some design will continue for a period of three months after construction commences. Construction will take 12 months. The total project period is 24 months.

Figure 5.9 (b), the fast-track expenditure plan, shows design and construction periods which are similar to those required for the traditional system. The total project period is 15 months.

The cost of the elements which comprise the project (site, design, construction, etc) are shown in the expenditure plans.

Site acquisition cost (the property itself and associated fees) and design cost are the same for both alternatives.

The construction cost shown in figure 5.9 (b) is estimated at £½ million more than that required when the traditional system is used. That sum may be regarded as an allowance for the additional cost (of construction management, accelerated progress, less competitive tendering and similar factors) above that which might be expected if the traditional system were used. Some contractors might question if the cost of construction is, in practice, higher if a fast-track approach is used.

A significant factor in the cost comparison is the timing of expenditure and the associated financing costs. The design team is to be paid as the design is developed. The contractor is to be paid monthly on the basis of the value of work executed. Financing charges follow those payments.

For purposes of the case study an interest rate of 12% pa has been used. Charges have been calculated quarterly.

To demonstrate the advantage of early completion on a commercial project figure 5.9 (b) shows a rental income for the period between the estimated completion date using a fast-track approach and the estimated completion date using the traditional system, namely, nine months.

It may be seen that although the construction cost of the fast-track approach is shown as £½ million more than that of the traditional system, at the end of the two year period, when the traditional system would produce a completed building, the fast-track approach shows a saving of £2 million over the traditional system.

In practice, the rent would not be shown as coming back as a saving. A trader developer would sell the development and so accrue a profit whereas an investment developer would re-negotiate medium term finance, say 25 years (similar to a mortgage) so that construction finance could be paid off and that from the rent derived he would be able to pay the interest on his medium term finance.

The other benefit to a developer is that by the building being completed sooner it can be valued as a complete development and shown in his accounts as an asset against which he can borrow further money to carry out further developments and grow more quickly.

5.4 Comparison of procurement systems

It should be emphasised that the systems as described are subject to greater variation.

Client's need The first component in each of the figures which has been used to illustrate the systems has read: 'client with need for a building' and the following assessments are concerned with that need. The client's viewpoint has been adopted for purposes of the comparison which follows.

Selecting the most appropriate procurement path is largely a matter of determining which performance requirements head the client's list of priorities. These might include:

(a) technical complexity
(b) aesthetics/prestige
(c) economy
(d) time
(e) exceptional size or complexity involving input from numerous sources and/or to satisfy several users' requirements
(f) price certainty at an early stage in the project's design development
(g) facility for the client to change/vary the works during the project's construction stage.

In figure 5.10 each of the requirements listed above is related insofar as it is able to satisfy the requirement.

Ratings have been given on a 1 to 5 scale with '1' the minimum and '5' the maximum capacity to meet the requirement. The ratings are the author's assessments of 'satisfaction'. It is assumed that the competence of the personnel involved is similar in all instances – only the systems are being compared. The following comparisons do not take into account all the characteristics of the systems which have been discussed.

Traditional system The traditional system rates '4' for projects with high technical complexity and/or with high aesthetic standards because the design team is not submitted to pressure, provided the design is essentially complete before competitive tenders are sought. In this event the team is able to develop the design rationally. It is the system which has provided the majority of designs in the past. Competitive tenders ensure that the client obtains the benefit of the lowest building cost. The system should produce a high rating for economy but the rating is reduced because the need to complete the design before commencing construction extends the overall project period. The interest paid by the client on the capital invested during the relatively long project period adds to the total cost of the project.

The sequential nature of the system and the experience gained by a significant number of clients of poor performance on exceptionally large or complex projects has

Client's performance requirements/expectations	traditional	management contracting/ construction management	package deal/design-and-build	project manager/client's representative
(a) technical complexity; the project has a high level of structural mechanical services or other complexity.	4	5	4	5
(b) high aesthetic or prestige requirements	5	3	3	4
(c) economy; a commercial or industrial project or project where minimum cost is required.	3	4	4	4
(d) time is of essence; early completion of the project is required.	2	4	5	4
(e) exceptional size and/or administrative complexity; involving varying client's/user requirements, political sensitivity etc.	2	4	4	5
(f) price certainty; is required at an early stage in the project's design development	4	2	4	4
(g) facility for change/variation control by client, users or others during the progress of the works.	5	5	1	4

5.10 *Rating the systems*

prompted the low ratings for performance requirements (d) and (e).

Price certainty should be capable of achievement provided the project has been fully designed and documented before tenders are sought. Priced bills of quantities facilitate the measurement and valuation of variations during the progress of the works. Hence the high ratings for requirements (f) and (g).

Management for a fee A rating of '5' has been given to requirement (a) because involvement of the construction team at an early stage in the development of the design should facilitate design of complex structures, mechanical services and other elements.

The system would not appear to offer advantages for requirement (b) and a median rating has been given.

Ratings of '4' have been given for performance requirements (c) and (d) because competition between management contractors, initially, and work-package contractors, subsequently, produce competition for building works. Because design and construction proceed in parallel the project period is kept to a minimum.

The participation of the management contractor as a member of the 'team' rather than as an outsider makes the system more satisfactory for exceptionally large or complex projects than the traditional system; hence the '4' rating for (e).

Price certainty at an early stage in the

project's development is not possible because the cost of building is not known until tenders have been accepted for all work packages. A correspondingly low rating has been given to requirement (f). That said, the client is normally provided with reasonably reliable estimates of the cost of the work packages by the consultant quantity surveyor as the design and construction develops.

A characteristic of these systems is parallel working. This makes it possible to vary the works until the work packages have been placed. Requirement (g) has been given a correspondingly high rating.

The advantages of these systems increase with the size of the project and extent to which time becomes the essence of the contract.

Package deal: design-and-build Performance requirements (a) and (c) have been given ratings of '4'. The involvement of designers with constructors (builders/contractors) on a team basis from the inception stage of the project should produce the expertise to cope with any technical complexity which the project may present. The result should be buildability – an unattractive word describing a necessary characteristic of any construction project which is to succeed. Economy should be achieved because the team is concerned not just with producing a design but with building to a budget. The discipline of designing and building should ensure that materials and components are selected at design stage which are economical and available. These remarks about economy apply also to time.

Components of building systems can be designed into the proposed building which will ensure that construction time is kept to a minimum. As the team has both design and construction organisation and control under one roof it can arrange that drawings and specifications are available as and when they are required to ensure that the regular progress of the works is not disrupted. Design and construction progress concurrently not consecutively, which minimises total project time. Ratings of '4' may be less than generous for these performance requirements as may '3' for requirement (b). A rating of '5' has been given to requirement (d).

Package-deal systems are used for large and complex projects such as nuclear power plants and petro-chemical developments. For projects of exceptional size or complexity the system should therefore be appropriate provided the 'contractor' ensures that a member of the firm who has the managerial expertise is appointed to 'stand outside' the day-to-day activities and hold a balance between design and construction interests. A high, but not the highest, rating has been given to this system in this respect because it is likely that not all firms would have the level of managerial expertise necessary to undertake projects of exceptional size or complexity.

A rating of '4' has been given to requirement (f), price certainty, because the price is normally agreed on the basis of the client's requirements. Provided the requirements are not changed after the contractor has submitted his proposals the price should hold. From this it follows, however, that there is little facility for cost control of changes during the progress of the works. The rating of '1' for requirement (g) reflects this.

Project management/client's representative Project management developed in response to demands for better management on exceptionally large and complex projects. However, there is increasing evidence that the system is gaining popularity among architects for more run-of-the-mill projects. This departure may at first sight appear surprising because when use of the system increased in the 1960s architects were its principal opponents. They saw their authority being undermined. One can only assume that this change of mind has been made because the experience of some architects on projects which have involved project managers has shown them some advantages of the system.

The term 'client's representative' was adopted by the British Property Federation in 1983 when devising the BPF system in preference to 'project manager'. The system, as such, has not been widely used by clients of the building industry but it is essentially a project manager/client's representative led design-and-build system and should be regarded as such when making a comparison of procurement systems.

A project manager may be employed with any of the systems described above to the advantage of all. Project management should improve the performance of all the systems because it enables the client, design team and construction team to concentrate their energies and skills on the functions for which they are primarily trained. A project manager should also promote teamwork among designers, builders and installers to the benefit of the client.

However, the greatest advantage should be a clear definition of the client's requirements through a brief to which the project manager is a principal contributor.

For these reasons the system has been rated highly. At the same time, a project manager cannot compensate for 'weaknesses' which are inherent in a system, such as the inability of the package-deal system to facilitate cost control of changes during the progress of the works.

Few people would argue with a high rating for project management where exceptional size or complexity are performance requirements.

For information regarding the incidence of use of the alternative procurement systems and clients' needs and expectations see sections 5 and 6 of *Building Procurement Systems* to which reference is made in *Sources* at the conclusion of this section.

Conclusions When making comparisons it is essential to compare 'like with like'. For this reason it is difficult to make valid comparisons of the alternative systems. Each system has been developed to meet particular client needs. There is no universal system. If one seeks the system which best meets the client's performance requirements in broad terms the ratings discussed above provide a guide for ranking.

5.5 Case study

To illustrate the method of rating a hypothetical project for a housing association, registered as a charity, which provides homes for the elderly may be used. The association requires an 'advanced care unit' to meet the needs of residents from its various homes who will undergo treatment as short-stay or out-patients. Operating theatres and specialist equipment will be required. The estimated cost of the unit is between £1 m and £1.5 m and it will be funded by the sale of investments and from a legacy. Outline planning approval has been given for the unit to be built in the spacious grounds of one of the association's homes which is listed as a building of architectural interest in a conservation area. An experienced member of the association's board of management will give his time, freely, to act as client's representative.

Which of the client's performance requirements should be given priority? The table in figure 5.10 has been used as a basis for appraisal.

The association requires a building which will satisfy aesthetic standards associated with a listed building and which are consistent with a conservation area (requirement (b)). The association may also wish to retain the facility to vary the works during their progress as medical technology for the elderly develops (requirement (g)).

The association owns the site so it will not incur site purchase costs, nor will it incur interest charges on the cash required to purchase the site. Time is not of the essence (requirement (d)).

Requirement (e), exceptional size and/or administrative complexity, is not relevant to this project.

The ratings indicate that a traditional approach should be the most appropriate for this project.

5.6 Exercises

1 *Evaluate* the use of negotiated contracts from the point of view of both the client and the contractors.

2   *Discuss* the factors which a consultant should consider when advising a client as to the most appropriate contractual arrangement for a project.

3   In 1984 the British Property Federation (BFT) introduced a new system of building management in respect of design and construction.
    Critically compare this system with the traditional appointment of designers and contractors and *discuss* the advantages which the BPF system confers upon the client.

4   In recent years many contractors have diversified their activities by offering to prospective clients alternative types of contractual arrangement.
    Discuss the extent to which the following arrangements have benefited the contractor and the client and identify the circumstances in which each may be used:
    (a) fee contracting
    (b) design-and-build contracts.

5   An industrial client, operating within a restricted budget, is considering a £1.20 m factory extension to cater for increased market demand.
    Outline planning permission has been obtained and the design brief and budget estimate agreed. A fundamental requirement is that the building must be completed within ten months. Any extended period could result in losses of up to £2,000 per week.
    Prepare a report for the client explaining the options available for tender procurement and contractual arrangements and make recommendations.

6   *Evaluate* the use of design and build contracts from the viewpoint of both client and contractor.

7   Using the table in figure 5.10 *appraise* the following projects and *suggest* which procurement path would be the most appropriate for each of the proposed projects.
    (a) Industrial development, for developer, comprising six similar but elegant buildings with shared office/reception block in prestigious development area.
       Estimated construction cost £4.5 m. Time is of the essence; green field site. Buildings will be fitted out by lessees subsequently.
    (b) City centre redevelopment – shared venture, local authority and pension fund. Mixed building types; office, civic, retail (supermarket and lock-ups), residential, sports centre, some refurbishment of warehousing.
       Estimated construction cost £150 m.
       Site mainly car park and empty buildings, few remaining leases fall in during next year or two but as much of site as possible to be developed as soon as possible.
    (c) Performing arts centre; main theatre, two theatre 'workshops', restaurant, etc, for Performing Arts Trust, financed by anonymous philanthropist offering £3 m, to be developed and run by Trust. Site near Weschester centre on bank of river Wes (15 metres of alluvial silt deposit on boulder clay), donated by Weschester CC. Estimated cost range of proposed development, £4 m – £5 m.
    (d) Developing country, eastern Europe/Asia requires energy for new coastal conurbation with planned population of 1m + involving tourism and light industry. Country does not have own oil.

5.7 Glossary

Collaboration contract   A variant of the negotiated approach where having established a price for the project, the client and contractor agree a sum to be included in the builder's price for the management of the construction phase. Part of this to be paid to the client in return for secondment of a senior member of the client's organisation to act as construction manager.

Competitive tender see **Traditional contracting**

Consortia A consortium is the grouping together of three or more organisations, generally of differing skills, with the objective of carrying out a specific project.

Continuity tender A continuity tender is similar to a serial tender. Contractors competitively tendering for a project are informed that given satisfactory performance, they will be awarded a similar project to follow on from the completion of the first and that the price for this will be negotiated, possibly using the prices of the original bill.

Construction management (CM) Construction management or CM is the term used in the USA to describe management contracting. (See also **Professional construction management**.)

Cost reimbursement see **Fixed fee/Prime cost contract**

Design-and-build Design-and-build or design-and-construct is where the contractor provides the design and construction under one contract.

Design-and-construct see **Design-and-build**

Fast-tracking Fast-tracking is a means of reducing project time by the overlapping of design and construction. Each trade's work commences as its plans and specifications are substantially completed.

Fixed fee/Prime cost contract Under this arrangement the contractor carries out the work for the payment of a prime cost (defined) and a fixed fee calculated in relation to the estimated amount of the prime cost.

Fixed price contract A fixed price contract may be a lump sum contract or a measurement contract based on fixed prices for units of specific work.

Joint venture A joint venture is the pooling of the assets and abilities of two or more firms for the purpose of accomplishing a specific goal and on the basis of sharing profits/losses.

Lump sum contract With a lump sum contract, the contractor agrees to perform the work for one fixed price, regardless of the ultimate cost.

Management contracting With management contracting the contractor works alongside the design and cost consultants, providing a construction management service on a number of professional bases. The management contractor does not undertake either design or direct construction work. The design requirements are met by letting each element of the construction to specialist sub-contractors.

Management fee see **Management contracting**

Negotiated contract In a negotiated contract the client selects one main contractor, at the outset, with whom to negotiate. In essence the arrangement is the same as that for a two-stage tender.

Package deal A package deal follows the same lines as design-build, with the contractor providing the design and construction under one contract, but there is often the implication that the building provided will be of a standardised or semi-standardised type.

Procurement Procurement is the amalgam of activities undertaken by the client to obtain a building.

Professional construction management PCM is a term used in the USA to describe an arrangement whereby the tasks of planning, design and construction are integrated by a project team comprising the owner, construction manager and the design organisation.

Professional construction manager (PCM) or **Construction manager** The PCM acts as a management contractor (UK) specialising in construction management within the professional construction management concept.

Project management Project management is concerned with the overall planning and co-ordination of a project from inception to completion aimed at meeting the client's requirements and ensuring completion on time, within cost and to required quality standards.

Separate contracts With separate contracts the client's professional adviser lets contracts for the work with a number of separate contractors. This arrangement was commonplace prior to the emergence of the general contractor.

Serial tender A serial tender is where a number of similar projects are awarded to a contractor, following a competitive tender on a master bill of quantities. This master bill forms a standing offer open for the client to accept for a number of contracts. Each contract is separate and the price for each calculated separately.

Target cost contract This form of cost-reimbursable contract involves the fixing of a cost either for the complete project or in respect of certain elements only, eg labour, or materials, or plant. If the final cost deviates from the target, the saving or excess is divided between client and contractor in pre-determined proportions.

Traditional contracting The traditional form of contracting is where the client appoints an architect or other professional to produce the design, select the contractor and to supervise the work through to completion. The contractor is selected on some basis of competition.

Turnkey A turnkey contract is one where the client has an agreement with one single administrative entity, who provides the design and construction under one contract, and frequently effects land acquisition, financing, leasing, etc.

Two-stage tender With a two-stage tender three or four contractors with appropriate experience are separately involved in detailed discussions with the client's professional advisers regarding all aspects of the project. Price competition is introduced through an approximate or notional bill or schedule of rates. Further selection criteria are then used to determine which contractor carries out the job.

5.8 Sources

The author's *Building procurement systems*, 2nd edition, 1990, published by the CIOB contains a collection of abstracts of published works concerned with the subject described in this section.

5.9 References

[1] *Survey of problems before the construction industries*, report prepared for the Minister of Works by Sir Harold Emmerson, 1962, HMSO.
[2] *The planning and management of contracts for building and civil engineering works*, report of the (Banwell) Committee, 1964, HMSO.
[3] NEDO, *The public client and the construction industries*, 1975, HMSO.
[4] NEDO, *Thinking about building*, 1985, HMSO.
[5] NEDO, *Faster building for commerce*, 1988, NEDO.
[6] *Personal communication*, Julian Vickery, Greycoat Group.

INDEX

Numbers in *italic* refer to illustrations

ACAS 133
ACA form of contract 196
Accounting Practice 1, 25
Acts of God 93
Administration, sub-contract 59
Advanced S Curve, The *see* TASC
Agendas 58
Ajudication 61, 133, 134, 142
Alpha Industrial development 118
All-in rates 19
All Risks Contract Works policy 123
Alternative Method of Management (AMM) 182 ff
Alternative methods of selection 29
Analysis of work 102
Antiquities 97
Arbitration Acts 134, 136, 137, 145, 146, 160
    Agreement 142
    Rules 137, 138, 140, 142, 145, 153
Arbitration 65, 133, 134, 137, 138
    advantages of 163
    disadvantages of 163
    flow chart 139
    joint 146
Arbitration Case Study: Lucifer v Barchester 163
    Award, The 173
    Closing addresses 170 ff
Arbitrator, inspection by 153
Arbitrator's fees and expensee 162
    directions 149
    powers 145 ff
    qualifications 144
Architect, access for 53
    exclusion of persons by 53
Architect's instructions 44, 45, 49, 54, 66, 109
Architects' Journal, The 10
Assessment procedure 25
Assignment 28
Attributable profits 75
Award, The 159 ff, 173
    enforcement of 136
    reasoned 137, 161

Baden-Hellard *135*
Bankers' references 24
Bankruptcy 65

Banwell 30
    Committee 30
    Report 177, 178
Barrett 75, 76, 77
Base Date 96
BCIS 10
BEC 42
    flow chart 3
    guide 3
Bills of quantities 9, 11, 19, 28, 29, 49, 100, 179
Bonds, contract guaranteed 125
BPF Manual 133
    system 133, 196, 201
BPIC 47, *47*, 51, *51*
Break-oven analysis *18*
    calculations 17
    method 17
Brief, The 3, 6, 7, 10
    for major projects 5
Briefing consultants 6, 116
British Property Federation (BPF) 191, 197
    system 191, 192, *193*, 200, 201, 202
Budget, definition 70
Budgetary control 71
    responsibility 71
Building 10, 67, *68*
Building Project Information Committee (BPIC) 47
Burden of proof 155

Case studies 117, 201
    Alpha Industrial development 118
    Hawthorne Home Scenario 117
Cases:
    Brodie v Cardiff Corporation (1919) 172
    Bryant v Birmingham Hospital Saturday Fund (1938) 172
    Calderbank v Calderbank (1976) 157
    Carnell Computer Technology Ltd v Unipart Group Ltd 154
    Clements v Clark (1880) 171
    Crosby and Sons Ltd v Portland UDS 106
    Ellis-Don Ltd v The Parking Authority of Toronto 111
    Finnegan Ltd v Sheffield City Council 111
    Lench Ltd v London Borough of Marton 106
    Lloyd v Milward (1895) 171
    Minter Ltd v Welsh Health Technical Services Organisation (1980) 112

Mitchell Construction Kinnear Moodie Group v East Anglia Regional Hospital Board 153
Pioneer Shopping Ltd v BTI Tioxide Ltd 161
Reese and Kirby Ltd v The Council of Swansea (1985) 112
Richards v May (1883) 172
Rush and Tompkins v GLC 156
Shaw and Horowitz Construction v Frank of Canada 111
Sutcliffe v Thackrah (1974) 171
Tate and Lyle v GLC 112
Wharf Properties Ltd v Cumine Associates (1988) 106
Whitehouse v Jordan (1981) 154
Whittal Builders Company Ltd v Chester-le-Street DC 111
Wraight v PHAT (Holdings) Ltd 112
Cash availability 11
CCPI system 50
Certificate of Completion of Making Good Defects 63
Chair, The 58, 59
Checklist, practical completion 62, 63
Checklists 63
   data for inclusion in the brief 3
   for meeting convenor 56
   for tasks prepared by regular clients 5
   local authority housing department 6, 7, 41
CIArb 144
CIOB: Code of Estimating Practice 11, 12, 18, 19, 20, 21, 22, 42
   Cost: value reconciliation 75
   Direct Membership examination xiii
Civil Evidence Act 154
Claim, preparing a 100, 114
   cost of 112
   credibility of 115
   events leading to 104
   presentation of 113
   statement of 105
Claimant, definition 142
Claims 127 ff
   points of 152
Clerk of works 53, 54, 189
Client's instructions 8
   need 198
Computer applications to budgetary management 71
Computer programs 46
Completion Date 89, 91, 93, 94, 98
Conciliation 132
Contract Bills 28, 43, 44, 49, 50, 52, 96, 98
   Drawings 28, 43, 44, 49, 50, 96, 98
Contract documents 44, 100
   custody of 48
   sum 44
Consultations 7

Consultants' fees 195
Contract, aspects of 26
   entering into 26
   privity of 28
Contractor determination 66
Contractor's analysis 102
Contractors' All Risks policy 123
   Indemnity policy 123
Contractural arrangements 177 ff
   relationships 27, 27, 28
Control Module, The 73
Co-ordinated project information (CPI) 46 ff
Cost ceiling 5, 7, 8, 10
Cost control, definition 70
Cost data cycle 23, 23, 24
Cost plan, elemental 67, 68
Cost planning 9, 10
   data 8, 10, 24, 67
   estimating 12, 19
   framework 8 ff
   objectives 8
   practice 10
   principles of 8
Cost policy, life cycle 7
Cost, units of 8, 10
Costs 108 ff
   drainage 9
   employment 107
   establishment 110
   hospital 10
   operational 76
   overhead 12, 13, 14, 15, 17
   standard 10
Cost: value, reconciling the 75 ff, 77
Counterclaims 138, 152, 158, 159
Court of Appeal 161
Courts 128, 133, 136
   County 131
   High 131, 136, 144, 146, 147, 153, 161
Cover prices 11
CPI documentation 47
Credit agencies 25
Crowter 155
'Cube' method 10

Damage to property 122
Data, sources of 10
Day work sheets 50
Defects 25
Defects Liability Period 63
Delays 89 ff
Descriptive schedules 48
Design-and-build 3, 48, 63, 185
   projects 189
   system 187
Design and management team, selection of 5
Design-led construction works 182

Design risk 8
Diplock, Lord 162
Direct costs 12
Discovery of antiquities 60
Disputes 131, 132, 142
   alternative approaches to 132
   procudures 138 ff
Disputants 132, 133, 142
   definition 142
Distortion Module, The 73
Documentary evidence 154
Documents, service of 148
   privileged 153
DOM 60, 61, 62, 147, 182, 189
Domestic sub-contractors 29, 30
Drainage costs 9
Drawing schedules 100
Drawings 79, 180
   arrangements for 47
   contract 28, 100
   recording revisions to *103*
   working 100

Earthquakes 93
EEC 26
Eichleay's formula 110
Emmerson, Sir Harold 177
Emmerson Report 177
Employer's determination 66
Employer's liability 123
   requirements 186, 187
Estimating 11 ff, 18
   data for 12
   methodical approach 12
Estimator's report and ajudication 23
Evidence 154
   as to opinion 154
   documentary 154
   hearsay 155
   of fact 154
   oral 154
   real 154
Execution of works 42 ff
Expenditure 67
Experts 117
Express terms 27
Extension of time 60, 90
   written 90

Facts, examining 101
   recording 100
Faster building for industry, report 196
Fast-track contractual arrangements 3, 182
   approach 197, 198
Fax 46, 148
Feasibility 3, 5, 6, 10
   studies 6

Finance Acts 55
Finance charges 112
Final account 63, 64, 65, 118
   certificate 43, 45, 63, 145, 172, 173
   selection 18
Financial reports, interim 76
Financial assessment 25
   status 24
Fine Arts Commission 7
Fire and Special Perils 123
Fisher 13, 14
Flow charts, BEC 3
   for selecting JCT form of contract 4
Force majeure 66, 72
Forecasting logic 72
Foreseeable losses 75

Gibson-Jarvie 161
Global approach 106
Goldwyn, Sam 27, 46
Group board 70
   communications 56
   decisions 56
   management methods 57
'Group think' 116

Hawker 161
Hawthorne Home development 78 ff
   Scenario 78, 117, 119
Hearing, The 117, 142, 159, 164
   procedure with 141
   procedure without 140
High Court 131, 162
   of Ontario 111
Hudson's formula 110, 111
Hospital 47
   costs 10

ICA's SSAP 75
ICE 131
Inception and feasibility 3
Indirect costs 12, 14
Income planning 67
   Tax 55
Inflation 109
Injury to persons or property 54
Inspection chambers 9
Inspection and tests 52
Insurance Act 1969, Employers' Liability 123
Insurance 121 ff
   against injury 54
   against loss or damage 126
   Employers' Liability 123, 127
   main types 122 ff
   motor 125, 127
   personal accident 122
   policies 124, 126, 127

practice 122
public liability 124, 125, 126
self 127
Insurable interest 122
Integrated cost management 70
Interest 158
Interim certificates 54
   payments 54
'Invitation to treat' 26, 27

JCT – Joint Contracts Tribunal 195
JCT 80 xiii, 24, 28, 29 ff
   Appendix 29, 42, 54, 89, 97, 147
   Fixed Fee Form of Prime Cost Contract 3
   Guide 41
   Standard form of Building Contract 2, 137, 179, 181, 186
   Standard form of Management Contract 185
   Standard form of Nominated Sub-contractor *32, 34, 35*
   Standard form of Nomination of Sub-contractor 39, *40*
   Works Contract Conditions 3

Kurtosis modules 73, 74, 75, *75*

Labour and plant costs 107
League of Nations 136
Legal liability 123
   employers 123
Legal representation 147
Liquidated damages 89
Liquidator 65
Litigation 133
Local amenities societies 7
Local authorities 6, 7
Local authority housing departments 7
Loss and expense 60, 98 ff
Loss or damage to property 122, 126
'Lump sum fee' 115

Management contract system *183*
Management involvement 105
Management role 18, 19
Management and supervisory costs 106
Market conditions 11
   factor 6
   research 5
Master programme 48
Master of Works 189
Mediation 132
Meetings 56 ff
   advantages of 59
   attendance at 58
   conducting 58
   convening 56
   for disputes 148

frequency of 58
   membership 59
   pre-meeting arrangement 57
   types of 57
Meteorological Office records 94, 100
Mini-trial 132
Minutes of meetings 59
   of site meetings 104
Modules
   control 73, 75
   distortion 73, 75
   Kurtosis variables 74, *74*, 75
Monitoring 75
Motor insurance 125
   policy 125
Moxley, Jenner and Partners 182

Named persons 29
   sub-contractors 29
National Association of Scaffolding Contractors 42
National Building Specification 47
NEDO formula 109
   report *Faster building for commerce* xiii, 189, 196
Nominated procedures 30
   alternative method 30, 39
   basic method 30
Nominated sub-contractors 29
Non-completion of job 60
Notification Date 142, 144, 148, 162
NSC/1 24, 30, 31, *32*, 33, *34*, *35*, 36, 39
NSC/2 31, 33, 36, 37, 147
NSC/3 31, 33, 36, 39, *40*
NSC/4 31, 35, 36, 38, 39, 49, 60, 61, 62, 90, 91, 92, 93, 147

Obligations, sub-contractor's 90
Offer, effects of costs 158
   timing of 158
Offers to settle 185 ff
   Calderbank 157, 158
   open 155
   sealed 156
Operational costs, recording 76
Oral evidence 154
Overhead costs 12, 13, 15, 17
   direct 12
   indirect 12
Overtime, non-productive 107

Package deal: design-and-build 185, 186, 188, 200
   system characteristics 185
Package dealer 188
Park 16
Partial possession by employer 84

INDEX

Patent rights 53
Pegs 52
Performing Arts Trust 202
'Person-in-charge', the contractor's 53
Pleadings 152
Practical completion of Works 62 ff
Pre-contract negotiations 24
    procedures 1
Pre-construction timetable 196, *196*
Preponderance of probability 155
Preselection 18, 19
Pre-tender data flow *14*
    procedures *13*
Price contingencies 8
Procedures with hearing 141
    without hearing 140
Procedures, uniformity of 42
Procurement path 1, 3, 5
    selection of 1
    systems 178 ff
Productivity, loss of 107
Professional indemnity insurance 124
Profit margin 15, *15*, 16, 17
    intuitive approach 15
Profits 75, 112
Project appreciation 19, *20*, 21
Project by contractor 42
    by project manager 41
Project management *190*
Project, mobilisation of 41
Public Liability insurance policy 124

Quality decisions 7
Quantity surveyor 29, 49, 50, 97, 98, 110
Quotations, obtaining 19

Ratings 198, 199, 200
Real evidence 154
Recorder Percival 111
Register of companies 24
Relevant events 92, 93, 94, 95, 96, 97
Report on management for commercial
    projects 1
Respondent, definition 142
Retention 63, 64
Revenue line 17
RIBA 144, 173
    plan of work 1, 3, 5, 67
RICS 144
Risk analysis 5
Royalties 53
Royce, Norman 132, 133, 164, 173

Scott Schedule 151, *151*
Scottish Building Contract Committee 146
Selection, factors affecting 24, 25
  alternative methods of 29

Society of Construction Arbitrators 132
Special case procedure 136, 137
Specific Perils 66, 94
Specification 7
Staff 114, 115, 116
Standard Accounting Practices 25
Standard form of contract 28, 29
Standard Method of Measurement (SMM) 9, 44,
    47, 163, 172
Statement of standard accounting practice
    (SSAP) 75
Statutory obligations 51
    requirements 51
Stephenson *139*
Strikes, off-site 94
Sub-contractor selection 24
Sub-contractors
    domestic 29
    named 29
    nominated 29
Subpoena ad testificandum 154
Stubb, Sir William 111

TASC 72, *72*, 73, *73*
    dependent variables *73*
    distortion module *73*
    Kurtosis variables 74, *74*
    software 74, 75
Tasks to be undertaken during outline proposals
    stage 6
Tender data 15, 100
Tender, active after submission 23
    decision to 11, 19, *20*
    designer-led 178, 179
    invitation to 11
    submission of 18
Tendering 11, 18
    efficiency 16, 17, *17*
    two-stage 182
Tenders
    from suppliers 10
    from specialist contractors 10
    from sub-contractors 10
    'idiot' 16
    record of *16*
Terminating the contract, alternative methods
    65
Time: cost case study *197*
    relationships 196
Time scale 6
Traditional client-architect-contractor 3
Trial 117
Turner 111

***U**berrina fides* 121
Underwriters 126, 127
Unit, composite 9

Unit, functional 10
Unit rates 19
Units of cost 8, 9, 10, 17
   finished work 71

Valuation rules 49
VAT 31, 55, 173
   Agreement 55, 142
Variations 49 ff, 75
   administering 50
   design 51
   identifying 50
   recording 50
   valuation of 50
Viability studies 5
Visits, supervisory 107
Volume, calculating 10
Vouchers 50

War damage 66
Wars 93
Weather conditions 93
Wilberforce, Lord 154
Without prejudice offers 155, 156, 157
Witness, expert role of 116 ff
Wood Report 189
Work units, finished 71
Workmanship 52
Working parties 59
Works
   by employer 55, 56
   by persons employed 55, 56
   carrying out 42
   engaged by employer 55, 56
   mobilisation of 24
   practical completion of 62 ff
Wraight 112